The Science and Politics of Global Climate Change
A Guide to the Debate

Why is the debate over climate change so confusing? Some say that there is clear evidence of an impending crisis, others that the evidence for climate change is weak. Some say that efforts to curb greenhouse gases will bankrupt us, others that we can solve the problem at manageable cost. In these arguments, both sides cannot be right. Reports in the media perpetuate the conflict: they invariably attempt to present both sides of the argument in a balanced manner. As a result, it is hard for non-specialists to sort out and evaluate the contending claims.

In this accessible primer, Dessler and Parson combine their expertise in atmospheric science and public policy to help scientists, policy makers, and the public sort through the conflicting claims in the climate-change debate. The authors explain how scientific and policy debates work, summarize present scientific knowledge and uncertainty about climate change, and discuss the available policy options. Along the way, they explain WHY the debate is so confusing.

Anyone with an interest in how science is used in policy debates will find this discussion illuminating. The book requires no specialized knowledge, but is accessible to any college-educated general reader who wants to make more sense of the climate-change debate. It can also be used as a textbook to explain the details of the climate-change debate, or as a resource for science students or working scientists, to explain how science is used in policy debates.

ANDREW E. DESSLER is an Associate Professor in the Department of Atmospheric Sciences at Texas A&M University. He received his Ph.D. in Chemistry from Harvard in 1994. He did postdoctoral work at NASA's Goddard Space Flight Center (1994–1996) and then spent nine years on the faculty of the University of Maryland (1996–2005). In 2000, he worked as a Senior Policy Analyst in the White House Office of Science and Technology Policy, where he collaborated with Ted Parson. Dessler's academic publications include one other book: *The Chemistry and Physics of Stratospheric Ozone* (Academic Press, 2000). He has also published extensively in the scientific literature on stratospheric ozone depletion and the physics of climate.

EDWARD A. PARSON is Professor of Law and Associate Professor of Natural Resources and Environment at the University of Michigan. Parson holds degrees in Physics from the University of Toronto and in Management Science from the University of British Columbia, and a Ph.D. in Public Policy from Harvard, where he spent ten years as a faculty member at the Kennedy School of Government. He served as leader of the 'Environmental Trends' Project for the Government of Canada and as editor of the resulting book, *Governing the Environment: Persistent Challenges, Uncertain Innovations*. His most recent book, *Protecting the Ozone Layer: Science and Strategy* (Oxford University Press, 2003), received the 2004 Harold and Margaret Sprout Award of the International Studies Association. Parson has served on the Committee on Human Dimensions of Global Change of the National Academy of Sciences, and on the Synthesis Team for the US National Assessment of Impacts of Climate Change. He has worked and consulted for various international bodies and for the governments of both Canada and the United States, including a period in the White House Office of Science and Technology Policy (OSTP) where he collaborated with Andrew Dessler. He has researched, published, and consulted extensively on issues of environmental policy, particularly its international dimensions; the political economy of regulation; the role of science and technology in public issues; and the analysis of negotiations, collective decisions, and conflicts.

The Science and Politics of Global Climate Change

A Guide to the Debate

ANDREW E. DESSLER
Department of Atmospheric Sciences,
Texas A&M University

EDWARD A. PARSON
Law School and School of Natural
Resources and Environment, University
of Michigan

CAMBRIDGE UNIVERSITY PRESS
Cambridge, New York, Melbourne, Madrid, Cape Town, Singapore, São Paulo

Cambridge University Press
The Edinburgh Building, Cambridge CB2 8RU, UK

Published in the United States of America by Cambridge University Press, New York

www.cambridge.org
Information on this title: www.cambridge.org/9780521831703

First published 2006
Fourth printing 2007

Printed in the United Kingdom at the University Press, Cambridge

A catalog record for this publication is available from the British Library

ISBN-13 978-0-521-83170-3 hardback
ISBN-13 978-0-521-53941-8 paperback

Contents

Preface

The Kyoto Protocol, the first international treaty to limit human contributions to global climate change, entered into force in February 2005. With this milestone, binding obligations to reduce the greenhouse-gas emissions that are contributing to global climate change came into effect for many of the world's industrial countries.

This event has also deepened pre-existing divisions among the world's nations that have been growing for nearly a decade. The most prominent division is between the majority of rich industrialized countries, led by the European Union and Japan, which have joined the Protocol, and the United States (joined only by Australia among the rich industrialized nations), which has rejected the Protocol as well as other proposals for near-term measures to limit greenhouse-gas emissions. Even among the nations that have joined Kyoto, there is great variation in the seriousness and timeliness of the emission-limiting measures they have adopted, and consequently in their likelihood of achieving the required reductions.

There is also a large division between the industrialized and the developing countries. The Kyoto Protocol only requires emission cuts by industrialized countries. Neither the Protocol nor the Framework Convention on Climate Change, an earlier treaty, provides any specific obligations for developing countries to limit their emissions. This has emerged as one of the sharpest points of controversy over the Protocol – a controversy that is particularly acute since the Protocol only controls industrialized-country emissions for the five-year period 2008–2012. In its present form, it includes no specific policies or obligations beyond 2012 for either industrialized or developing countries. While the Kyoto Protocol represents a modest first step toward a concrete response to climate change, there has been essentially no progress in negotiating the larger, longer-term changes that will be required to slow, stop, or reverse any human-induced climate changes that are occurring.

As these political divisions have grown sharper, public arguments concerning what we know about climate change have also grown more heated. Climate change may well be the most contentious environmental issue that we have yet seen. Follow the issue in the news or in policy debates and you will see arguments over whether or not the climate is changing, whether or not human activities are causing it to change, how much and how fast it is going to change in the future, how big and how serious the impacts will be, and what can be done – at what cost – to slow or stop it. These arguments are intense because the stakes are high. But what is puzzling, indeed troubling, about these arguments is that they include bitter public disagreements, between political figures and commentators and also between scientists, over points that would appear to be straightforward questions of scientific knowledge.

In this book, we try to clarify both the scientific and the policy arguments now being waged over climate change. We first consider the atmospheric-science issues that form the core of the climate-change science debate. We review present scientific knowledge and uncertainty about climate change and the way this knowledge is used in public and policy debate, and examine the interactions between political and scientific debate – in effect, to ask how can the climate-change debate be so contentious and so confusing, when so many of the participants say that they are basing their arguments on scientific knowledge.

We then broaden our focus, to consider the potential impacts of climate change, and the available responses – both in terms of technological options that might be developed or deployed, and in terms of policies that might be adopted. For these areas as for climate science, we review present knowledge and discuss its implications for action and how it is being used in public and policy debate. Finally, we pull these strands of scientific, technical, economic, and political argument together to present an outline of a path forward out of the present deadlock.

The book is aimed at an educated but non-specialist audience. A course or two in physics, chemistry, or Earth science might make you a little more comfortable with the exposition, but is not necessary. We assume no specific prior knowledge except the ability to read a graph. The book is suitable to support a detailed case-study of climate change in college courses on environmental policy or science and public policy. It should also be useful for scientists seeking to understand how science is used – and misused – in policy debates.

Many people have helped this project come to fruition. Helpful comments on the manuscript have been provided by David Ballon, Steve Porter, Mark Shahinian, and Scott Siff, as well as seminar participants at the University of British Columbia, the University of Michigan School of Public Health, and the University of Michigan Law School. A. E. D. received support for this project from a NASA New Investigator Program grant to the University of Maryland, as well as from the University of Maryland's Department of Meteorology and College of Computer, Mathematical, and Physical Sciences. All these contributions are gratefully acknowledged. A. E. D. especially notes the contributions of Professor David Dessler, for discussions in which many of the early ideas for the book were developed or refined.

1

Global climate change: a new type of environmental problem

Of all the environmental issues that have emerged in the past few decades, global climate change is the most serious, and the most difficult to manage. It is the most serious because of the severity of harms that it might bring. Many aspects of human society and well being – where we live, how we build, how we move around, how we earn our livings, and what we do for recreation – still depend on a relatively benign range of climatic conditions, even though this dependence has been reduced and obscured in modern industrial societies by their wealth and technology. We can see this dependence on climate in the economic harms and human suffering caused by the climate variations of the past century, such as the "El Niño" cycle and the multi-year droughts that occur in western North America every few decades. Climate changes projected for the twentyfirst century are much larger than these twentieth-century variations, and their human impacts are likely to be correspondingly greater.

Projections of twentyfirst-century climate change are uncertain, of course. We will have much to say about scientific uncertainty and its use in policy debates, but one central fact about uncertainty is that it cuts both ways. If projected twentyfirst-century climate change is uncertain, then the actual changes might turn out to be smaller than we now project, or larger. Uncertainty about how the climate will actually change consequently makes the issue more serious, not less. Present projections of twentyfirst-century climate change include, at the upper end of the range of uncertainty, sustained rapid changes that appear to have few precedents in the history of the Earth, and whose impacts on human well-being and society could be of catastrophic proportions.

Climate does not just affect people directly: it also affects all other environmental and ecological processes, including many that we might not recognize as related to climate. Consequently, large or rapid climate change will represent an added threat to other environmental issues such as air and water quality, endangered ecosystems and biodiversity, and threats to coastal zones, wetlands, and the stratospheric ozone layer.

In addition to being the most serious environmental problem we have yet faced, climate change will also be the most difficult to manage. Environmental issues often carry difficult tradeoffs and political conflicts, because solving them requires limiting some economically productive activity or technology that is causing unintended environmental harm. Such changes are costly and generate opposition. But for the issues we have faced previously, technological advances and intelligent policies have allowed great reductions in environmental harms at modest cost and disruption, so these tradeoffs and conflicts have turned out to be quite manageable. Controlling the sulfur emissions that contribute to acid rain in the United States of America provides a good example. When coal containing high levels of sulfur is burned, sulfur dioxide (SO_2) in the smoke makes the rain that falls downwind of the smokestack acidic, harming lakes, soils, and forests. Over the past 20 years, a combination of advances in technologies to remove sulfur from smokestack gases, and well-designed policies that give incentives to adopt these technologies, burn lower-sulfur coal, or switch to other fuels, have brought large reductions in sulfur emissions at a relatively small cost and with no disruption to electrical supply.

Climate change will be harder to address because the activities causing it – mainly burning fossil fuels for energy – are a more essential foundation of world economies, and are less amenable to any simple technological corrective, than the causes of other environmental problems. Fossil fuels provide nearly 80 percent of world energy supply, and no technological alternatives are presently available that could replace this huge energy source quickly or cheaply. Consequently, climate change carries higher stakes than other environmental issues, both in the severity of potential harms if the changes go unchecked, and in the apparent cost and difficulty of reducing the changes. In this sense, climate change is the first of a new generation of harder environmental problems that human society will face over this century, as the increasing scale of our activities puts pressure on ever more basic planetary-scale processes.

When policy issues have high stakes, it is typical for policy debates to be contentious. Because the potential risks of climate change are so serious, and the fossil fuels that contribute to it are so important to the world economy, we would expect to hear strong opposing views over what to do about climate change – and we do. But even given the issue's high stakes, the number and intensity of contradictory claims advanced about climate change is extreme. The following published statements give a sense of the range of views about climate change.

From former US Vice-President Al Gore:

[T]he vast majority of the most respected environmental scientists from all over the world have sounded a clear and urgent alarm ... [T]hese scientists are telling the people of every nation that global warming caused by human activities is becoming a serious threat to our common future ... I don't think there is any longer a credible basis for doubting that the earth's atmosphere is heating up because of global warming ... So the evidence is overwhelming and undeniable. Global warming is real. It is happening already and the anticipated consequences are unacceptable.[1]

From former US Secretary of Defense and of Energy James Schlesinger:

What we know for sure is quite limited ... We know that the theory that increasing concentrations of greenhouse gases like carbon dioxide will lead to further warming is at least an oversimplification. It is inconsistent with the fact that satellite measurements over 24 years show no significant warming in the lower atmosphere, which is an essential part of the global-warming theory.[2]

From US Senator James Inhofe:

[A]nyone who pays even cursory attention to the issue understands that scientists vigorously disagree over whether human activities are responsible for global warming, or whether those activities will precipitate natural disasters ... So what have we learned from the scientists and economists I've talked about today?

1   The claim that global warming is caused by man-made emissions is simply untrue and not based on sound science.
2   CO_2 does not cause catastrophic disasters – actually it would be beneficial to our environment and our economy ...

With all of the hysteria, all of the fear, all of the phony science, could it be that man-made global warming is the greatest hoax ever perpetrated on the American people? It sure sounds like it.[3]

From the *Wall Street Journal*:

... the science on which Kyoto is based has never been able to explain basic questions. Most glaring is why the Earth warmed so much in the

---

[1] Global Warming and the Environment, speech by Al Gore, Beacon Hotel, New York City, Jan. 15, 2004.

[2] Commentary: Cold facts on global warming, James Schlesinger, *Los Angeles Times*, Jan. 22, 2004, p. B17.

[3] The Science of Climate Change, floor statement by Senator James M. Inhofe, July 28, 2003.

early part of the 20th century, before the boom in carbon dioxide emissions. Another is why the near-earth atmosphere (measured by satellites) isn't warming as much as the Earth's surface. There's also the nagging problem that temperatures more than 1,000 years ago appear to have been as warm, if not warmer, than today's.[4]

From the *National Post of Canada*:

Global warming, as increasing numbers of actual scientists will tell you, is not happening.[5]

From the well-known scientific skeptic, S. Fred Singer:

[T]he Earth's climate has not warmed appreciably in the past two decades, and probably not since about 1940.[6]

That the climate is currently warming rests solely on surface thermometer data. It is contradicted by superior observations from weather satellites and independent radiosonde data from weather balloons. Proxy (non-thermometer) data from tree rings, ice cores, etc., all confirm that there is no current warming. That the 20th century was the warmest in the past 1,000 years derives entirely from misuse of such proxy data. . . . The claim that climate models . . . accurately reproduce the temperature record of the past 100 years, is spurious.[7]

From Nobel laureate F. Sherwood Rowland, of the University of California at Irvine:

The earth's climate is changing, in large part because of the activities of humankind. The simplest measure of this change is the average temperature of the Earth's surface, which has risen approximately 0.7 degrees Celsius over the past century, with most of this increase occurring in the past two decades. In other words, the Earth is undergoing global warming . . . The possibility exists for notable deterioration of the climate in the United States even on a decadal time scale . . . [T]he climate change problem will be much more serious by the year 2050 and even more so by 2100.[8]

---

[4] Global warming glasnost, editorial, *Wall Street Journal*, Dec. 4, 2003, p. A16.

[5] The Conservatives must attack Kyoto, editorial, *National Post of Canada*, March 19, 2004.

[6] S. Fred Singer, testimony before the US Senate Committee on Commerce, Science, and Transportation, July 18, 2000.

[7] S. Fred Singer, Bad data make global warming a cold case, letter to the editor, *Wall Street Journal*, Nov. 10, 2003, p. A17.

[8] F. Sherwood Rowland, Climate change and its consequences: issues for the new U.S. Administration, *Environment* **43**(2), March 2001, pp. 29–34.

And from Jerry Mahlman, former director of the US Geophysical Fluid Dynamics Laboratory at Princeton:

> ... we know that the earth's climate has been heating up over the past century. This is happening in the atmosphere, ocean and on land ... [I]f the climate model projections on the level of warming are right, sea level will be rising for the next thousand years, the glaciers will be melting faster and dramatic increases in the intensity in rainfall rates and hurricanes are expected ... Unfortunately, these projections are based on strong science that refuses to go away. Oh sure, there are people insisting that warming is just a part of natural weather cycles, but their claims are not close to being scientifically credible ... These people remind me of the folks who kept trying to cast doubt on the science linking cancer to tobacco use. In both situations, the underlying scientific knowledge was quite well established, while the uncertainties were never enough to render the problem inconsequential. Yet, this offered misguided incentives to dismiss a danger ... Global warming is unpleasant news. The costs of doing something substantial to arrest it are daunting, but the consequences of not doing anything are staggering.[9]

One of the most striking aspects of this debate is the intensity of disagreements expressed over what we might expect to be simple matters of scientific knowledge, such as whether the Earth is warming or not. Such heated public confrontation over the state of scientific knowledge and uncertainty – not just between political figures and policy commentators, but also between scientists – understandably leaves most concerned citizens confused. The state of public and political debate on the issue makes it hard for non-specialists to understand what the advocates are arguing about, or to judge the strength of competing arguments.

Our goal in this book is to clarify the climate-change debate. We seek to help the concerned, non-expert citizen to understand what is known about climate change, and how confidently it is known, in order to develop an informed opinion of what should be done about the issue. We will summarize the state of knowledge and uncertainty on key points of climate science, and examine how some of the prominent claims being advanced in the policy debate – including some in the quotes above – stand up in light of present knowledge. Can we confidently state that some of these claims are simply right and others simply wrong, or are these points of genuine uncertainty or legitimate differences of interpretation?

[9] Claudia Dreifus, A Conversation with Jerry Mahlman: listening to climate models and trying to wake up the world, *New York Times*, Dec. 16, 2003, p. F2,

We will also summarize present understanding and debate over the likely impacts of climate change and the responses available to deal with the issue – matters that go beyond purely scientific questions, but which can be informed by scientific knowledge.

We will also examine how scientific argument and political controversy interact. This will help to illuminate why seemingly scientific arguments play such a conspicuous role in the climate-change policy debate, and in particular how such extreme disagreements can arise on points that would appear to be matters of scientific knowledge. What do policy advocates hope to achieve by arguing in public over scientific points, when most of them – like most citizens – lack the knowledge and training to evaluate these claims? Why do senior political figures appear to disagree on basic scientific questions when they have ready access to scientific experts and advisors to clarify these for them? And finally, what are the effects of such blended scientific and political arguments on the policy-making process?

While there is plenty of room for honest, well-informed disagreement over what should be done about global climate change, it is our view that the issue is made vastly more confused and contentious than it need be by misrepresentations of the state of scientific knowledge in policy debate – in particular, by exaggeration of the extent and significance of scientific uncertainty on key points about the global climate and how it might respond to further human influences.

Before we can engage these questions, the next two sections of this chapter provide some necessary background. Section 1.1 provides a brief background on the Earth's climate and the basic mechanisms that control it and can change it. Section 1.2 provides a brief history of existing policy and institutions concerned with global climate change, to provide the policy context for the present debate.

1.1   Background on climate and climate change

The climate of a place, a region, or the Earth as a whole, is the average over time of the meteorological conditions that occur there – in other words, its average weather. For example, in the month of November between 1971 and 2000 in Washington D.C., the average daily high temperature was 14 °C, the average daily low was 1 °C, and 0.3 cm of precipitation fell.[10] These average values, along with averages of other meteorological quantities such as humidity, wind speed, cloudiness, and snow and ice coverage, define the November climate of Washington over this period. While climate refers to average meteorological conditions, weather refers to meteorological conditions at a particular time. For example, on

---

[10] Data from the NOAA National Climatic Data Center web page: http://lwf.ncdc.noaa.gov/oa/climate/climateresources.html

November 29, 1999, in Washington, D.C., the high temperature was 5 °C, the low was −3 °C, and no precipitation fell. The weather on this particular November day in Washington was somewhat colder and drier than Washington's average November climate.

Weather matters for short-term, day-to-day decisions. Should you take an umbrella when you go out tomorrow? Will freezing temperatures kill plants left outdoors tonight? Is this a good weekend to go skiing in the mountains? Should you move your outdoor party scheduled for this weekend indoors? In each of these cases, you do not care about long-term average conditions, but about conditions at a specific time – not the climate, but the weather.

Climate matters for longer-term decisions. If you run an electric utility, you care about the climate because if average summer temperatures increase, people will run their air conditioners longer each day and consume more electricity. In this case, you may need to build more generating plants to meet this increased demand. If you are a city official, you care about the climate because urban water supplies usually come from reservoirs fed by rain or snow. Changes in the average temperature or in the timing or amount of precipitation could change both the supply and the demand for water. Consequently, if the climate changes, the city may need to expand capacity to store or transport water, find new supplies, or develop policies to limit water use in times of scarcity.

To understand the processes that are changing the climate, we must first consider why the climate is the way it is, in particular places and for the Earth as a whole. Scientists have been studying these questions since the early nineteenth century, beginning with the largest question of all: why is the Earth the temperature that it is? The Earth is warmed by the Sun and cooled by emitting radiation to space. The Earth's temperature is determined by the relationship between the incoming radiation the Earth absorbs from sunlight and the radiation it emits back to space. Not all the sunlight that strikes the Earth is absorbed, however. About 30 percent is reflected back into space – which is why the Earth looks bright when viewed from space – while the other 70 percent is absorbed and warms the surface and lower atmosphere. For the Earth to stay at a constant temperature, the total energy of the incoming and outgoing radiation must be equal. Because the Sun is so hot (about 5400 °C), sunlight is strongest in the visible and near-infrared region of the electromagnetic spectrum (with wavelengths from about 0.4 to 1 micron). The Earth is much cooler, so the radiation it emits is of longer wavelengths, lying in the infrared region (with wavelengths from about 5 to 20 microns). This is the region of the electromagnetic spectrum that certain types of night-vision goggles use to give clear images in total darkness, detecting minor temperature differences among objects and people by the infrared radiation they emit. A simple calculation can determine what the average temperature of the Earth should be for the outgoing radiation just to balance the energy of the absorbed sunlight. This calculation indicates that the average temperature of the Earth's surface should be about $-20\,°C$.

This is awfully cold. Fortunately, it is also wrong. The Earth's surface is much warmer than this, a pleasant $15\,°C$ on average. The error in the calculation comes from assuming that the infrared radiation emitted from the Earth passes directly to space. It does not, because it must pass through the atmosphere. And while the air in a clear sky is nearly transparent to the visible wavelengths coming in from sunlight, air absorbs the infrared radiation emitted by the Earth fairly strongly. This absorption is not caused by the main components of the atmosphere, molecular nitrogen and oxygen: these gases are as transparent to infrared radiation as they are to visible light. Rather, the absorption comes from several minor atmospheric constituents, principally water vapor and carbon dioxide (CO_2). By absorbing and re-emitting infrared radiation throughout the atmosphere, these gases impede the passage of radiation from the Earth's surface to space. This process warms the Earth's surface and lowest ten kilometers of the atmosphere, while cooling the atmosphere at higher altitudes. Ever since this natural warming mechanism was first described in the nineteenth century, it has been widely called the "greenhouse effect." More recently, it has been compared to wrapping a blanket around the Earth. Neither of these analogies is really accurate, however, since both blankets and greenhouses mainly work by slowing the physical escape of warm air rather than by disrupting the passage of radiation.

The power of these "greenhouse gases" to warm the Earth's surface is awesome. Although these gases are present in the atmosphere at only minute concentrations, they warm the surface by nearly $35\,°C$. Their power becomes even clearer if we compare the climate of the Earth to that of the neighboring planets, Mars and Venus. Mars has a thin atmosphere that is almost completely transparent to infrared radiation, giving it an average surface temperature below $-50\,°C$. Venus has a dense, CO_2-rich atmosphere that produces an intense greenhouse effect, raising its average surface temperature above $450\,°C$ – hot enough to melt lead.

But if greenhouse gases in the atmosphere warm the Earth to its present habitable state, increasing the concentration of these gases could make the Earth warmer still. This possibility was proposed by the Swedish chemist Svante Arrhenius in 1906, and again with more supporting evidence by the British engineer Guy Callendar in 1938. These proposals were not initially taken seriously, because with the crude tools then available to observe infrared radiation, it looked like the levels of CO_2 and water vapor already in the atmosphere were absorbing enough radiation to create the maximum possible greenhouse effect. By the 1950s, however, more precise measurements of infrared spectra showed this belief to be

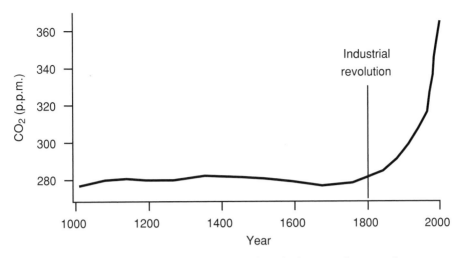

Figure 1.1. Global average concentration of CO_2 in the atmosphere over the past 1000 years, in parts per million (p.p.m.). Source: Figure SPM-2, IPCC (2001a).

wrong, so increasing CO_2 could increase infrared absorption in the atmosphere and raise the surface temperature.

CO_2 is not the only greenhouse gas, nor is it the only one emitted by human activities. Other greenhouse gases that are increasing due to human activities include: methane (CH_4), which is emitted from rice cultivation, livestock, biomass burning, and landfills; nitrous oxide (N_2O), which is emitted from various agricultural and industrial processes; and the halocarbons, a group of synthetic chemicals of which the most important are the chlorofluorocarbons (CFCs), which are used as refrigerants, solvents, and in various other industrial applications. Human activities do not control all greenhouse gases, however. The most powerful greenhouse gas in the atmosphere is water vapor. Human activities have little direct control over its atmospheric abundance, which is controlled instead by the worldwide balance between evaporation from the oceans and precipitation.

By the 1950s and early 1960s, it was also becoming clear that human activities were releasing CO_2 fast enough to significantly increase its atmospheric abundance. Figure 1.1 shows how the abundance of CO_2 in the atmosphere has varied over the past 1000 years – remaining nearly constant for most of the millennium, then beginning a rapid increase around 1800. This rapid increase closely tracked the sharp rise in fossil-fuel use that began with the industrial revolution.

Despite clear evidence of increasing atmospheric CO_2, during the 1960s and 1970s scientific views about likely future climate trends were divided. Some scientists expected the Earth to warm from rising concentrations of CO_2 and other greenhouse gases. Others expected the Earth to cool, based partly on the record of past climate oscillations between ice ages and warm interglacial periods. The present warm period has lasted about 10 000 years, roughly the same length as previous interglacial warm periods, suggesting that we might be due for a gradual, long-term cooling as we head into another ice age. Moreover, global temperature records between about 1945 and 1975 showed a slight cooling trend. It was also clear that smoke and dust emitted by human activities could shade the Earth's surface from incoming sunlight and so magnify any natural cooling trend. By the early 1980s, however, global temperatures had resumed warming and many new pieces of evidence indicated that greenhouse gases were the predominant human influence and that warming was the predominant direction of concern.

As we will discuss in Chapter 3, the best present projections are that if emissions of CO_2 and other greenhouse gases keep growing more or less as they have been, by the end of the twentyfirst century the Earth's average temperature will rise by a few degrees Celsius. This increase might not sound like much, since many places on Earth experience much larger temperature swings. The difference between a hot summer day and a cold winter one can be as large as 50 °C, and changes half that large can occur from day to night or from one day to the next. Therefore, you might reasonably guess that an increase of a few degrees in the global temperature is not likely to matter much. But there is a serious error in this line of reasoning. While the temperature of any single place on the Earth can vary greatly, the average temperature of the whole Earth is quite constant, throughout the year and from year to year. In the Earth's past, changes of only a few degrees in global-average temperature have been associated with extreme changes in climate. For example, at the peak of the last ice age 20 000 years ago – when glaciers thousands of feet thick covered most of North America – the average temperature of the Earth was only about 5 °C cooler than it is today. Thus, the prospect of a few degrees Celsius rise in global temperature over just 100 years – and perhaps more beyond – must be considered with the utmost seriousness. In Chapter 3 we will summarize what has been learned since climate change emerged as a serious scientific question nearly 50 years ago, about the evidence for present changes, likely future changes, and their impacts.

Aside: climate change and ozone depletion

People frequently confuse global climate change with depletion of the stratospheric ozone layer, but these are two distinct environmental problems. Ozone is a molecule made up of three oxygen atoms, which occurs naturally in the stratosphere (the atmospheric region from about 15 to 40 kilometers above the surface). Ozone in the stratosphere protects life on

Earth by absorbing most of the highest-energy ultraviolet (UV) radiation in sunlight. To make things more confusing, ozone in the lower atmosphere (the troposphere) is a health hazard and a major component of smog, which human activities are increasing. To keep "good ozone" (up there) and "bad ozone" (down here) straight, simply remember that you want ozone between you and the Sun, but do not want to breathe it.

Beginning in the 1970s, scientists realized that a group of manmade chemicals, of which the most important were the chlorofluorocarbons or CFCs, could destroy ozone in the stratosphere. The result would be more intense UV radiation reaching the surface, causing an increase in skin cancer, cataracts, and other harms to human health and ecosystems. Concern mounted further in the 1980s, when extreme ozone losses were observed over Antarctica every spring (October and November) – the "ozone hole" – and CFCs were identified as the cause.

After ten years of unsuccessful attempts to solve the problem, nations agreed in the late 1980s and 1990s to a series of strict regulatory controls that have now nearly eliminated most ozone-depleting chemicals in the industrialized countries. Developing countries are now moving toward phasing out the same chemicals. Because of these controls, the concentration of CFCs in the atmosphere has already begun to decline, and stratospheric ozone is projected to recover gradually over the next 30 to 50 years.

There are a few ways that climate change and ozone depletion are linked. One connection is that CFCs are strong absorbers of infrared radiation, so they contribute to climate change as well as destroying ozone. Another connection is that while climate change warms the Earth's surface and lower atmosphere, it will also make the stratosphere colder and wetter. Colder and wetter conditions are more favorable for ozone destruction, and so are likely to delay the recovery of the ozone layer even if worldwide reductions of ozone-depleting chemicals stay on course. But despite these linkages, ozone depletion and climate change are fundamentally different environmental problems. They have different causes: CFCs and certain other chemicals containing chlorine or bromine, versus CO_2 and other greenhouse gases. And they have different effects: more intense UV radiation reaching the Earth's surface, harming health and ecosystems, versus changes in climate and weather worldwide. Although there are important differences between the two issues, many aspects of how nations responded to ozone provide useful analogies or lessons for how to respond to global climate change. Consequently, we will refer to specific relevant aspects of the ozone issue at several points throughout this book.

1.2   Background on climate-change policy

Like many serious environmental issues, global climate change came to the attention of policy-makers after decades of related scientific research. Climate change attracted virtually no public or political attention in the 1960s, and only a little during the energy-policy debates of the 1970s. By this time it was clear that human activities had the potential to change the global climate, but it was not yet clear whether the predominant direction of human influence would be warming or cooling. But by the early 1980s, as it became increasingly clear that warming from greenhouse gases was the predominant concern, scientists and scientific organizations began trying to persuade governments to pay attention to the climate problem. They had little success until 1988, when several events brought climate change suddenly to the top of the political agenda.

That summer, North America suffered an extreme heat wave and the worst drought since the dust-bowl years of the 1930s. By July, 45 percent of the United States was in a drought and a few prominent scientists stated publicly that global climate change was probably the cause. Moreover, this extreme summer followed a period of intense worldwide publicity about the Antarctic ozone hole and the negotiation of the Montreal Protocol, the international treaty to control the responsible chemicals. Under these conditions, politicians and the public were primed to consider the possibility that human activities could be disrupting the global climate. In late 1988, instead of naming a "Person of the Year", *Time Magazine* designated "Endangered Earth" the "Planet of the Year," while the United Nations General Assembly passed a resolution stating that the climate was "a concern to mankind."

Governments' first response was to establish an international body to conduct assessments of scientific knowledge of climate change, the Intergovernmental Panel on Climate Change or IPCC. The IPCC involved hundreds of scientists organized into three working groups, each responsible for a different aspect of the climate issue: the atmospheric science of climate change; the potential impacts of climate change and ways to adapt to the changes; and the potential to reduce the greenhouse-gas emissions contributing to climate change. The three major assessment reports that the IPCC has completed since its formation, in 1990, 1995, and 2001, are widely regarded as the authoritative statements of scientific knowledge about climate change. We will refer to the conclusions of these assessments repeatedly throughout this book.

As the IPCC was beginning its work in the late 1980s, governments also began considering concrete measures to respond to climate change. Over the two years following the hot summer of 1988, several high-profile international political conferences called for reducing worldwide CO_2 emissions, typically by 10 to 20 percent as a first step. Through 1991 and 1992, national representatives worked to negotiate the first international treaty on climate change, the Framework Convention on Climate Change (FCCC). Signed in June 1992, this treaty entered into force in 1994 and has since been established law in all the nations that have ratified – now numbering nearly 190, including the United States.[11]

The FCCC's stated objective is "Stabilization of greenhouse gas concentrations in the atmosphere at a level that would prevent dangerous anthropogenic interference with the climate system . . . within a time-frame sufficient to allow ecosystems to adapt naturally to climate change, to ensure that food production is not threatened, and to enable economic development to proceed in a sustainable manner." The treaty also states several principles intended to guide subsequent climate-policy decisions, of which a particularly important one is the principle of "Common but differentiated responsibility." This principle states that all nations have an obligation to address the climate issue, but not in the same way or at the same time, and in particular that ". . . the developed-country Parties should take the lead in combating climate change and the adverse effects thereof."[12]

The FCCC was not intended to be the final word on the climate issue, but to provide a starting point for more specific and binding measures to be negotiated later. Consequently, in contrast to its ambitious principles and objectives, the treaty's concrete measures were weak and preliminary. Under the FCCC, parties committed to reporting their current and projected national emissions and supporting climate-related research. In addition, all parties undertook a general obligation to take measures to limit emissions and report on these. What these measures had to be, or had to achieve, however, was not specified. Only for the industrialized countries (or "Annex 1 countries") did this general obligation also include the specific aim of returning emissions to 1990 levels by 2000. This aim was the closest the FCCC came to concrete action to advance its objectives, but even it was not legally binding.

Weak as this aim was, few governments made serious efforts to meet it. Many, including the USA, assembled national programs that were little more than exhortations for voluntary action and re-labelings of existing programs. The few nations that met the emission-reduction target largely did so by historical accident or through policies adopted for other reasons. Russia, for example, met its target because of the collapse of the Soviet economy after 1990, Germany because it

---

[11] After a treaty has been negotiated and signed by national representatives, it enters into force, or becomes legally binding, only after enough nations take the second step of ratifying it – formally expressing their commitment to be bound by it. Every treaty specifies how many nations must ratify for it to enter into force. After these are received, the treaty becomes binding upon those who have ratified.

[12] Framework Convention on Climate Change, Article 3.1 (available at http://unfccc.int/resource/docs/convkp/conveng.pdf).

absorbed the shrinking East German economy, and Britain because it was privatizing electrical generation and cutting subsidies for coal production.

It was clear immediately after adoption of the FCCC that achieving significant emission reductions would take stronger measures. After a few years of wide-ranging debate about various forms these stronger measures might take, discussions shifted by 1995 to negotiating binding national greenhouse-gas emission limits. These negotiations culminated in the signing of the Kyoto Protocol in December 1997.[13]

Negotiations of the Kyoto Protocol were marked by hard, last-minute bargaining over the distribution of national limits. European and Japanese delegations sought stringent cuts, by 5 to 15 percent below 1990 levels by 2010. The Clinton administration was initially reluctant to accept near-term emission cuts, and instead proposed only research and voluntary initiatives in the early years, with emission limits coming into effect only after 2008. The US Senate took the unusual step of expressing its hostility to emission limits before negotiations were completed, by passing a resolution that rejected new emission commitments for industrial countries unless developing countries also cut emissions at the same time.

The agreement reached in the final hours of the Kyoto Conference imposed specific emission-reduction targets for each industrialized country over a five-year "commitment period" of 2008–2012. Targets were defined for total emissions of a basket of CO_2 and five other greenhouse gases. Despite the Senate resolution, the US delegation signed the treaty even though it included no emission limits for developing countries. The required emission reductions were 8 percent for the European Union and a few other European nations; 7 percent for the United States; 6 percent for Japan and Canada; and zero (i.e. hold emissions at their baseline level) for Russia and Ukraine.[14] If all nations met their targets, the total emission reduction from these nations would be 5.2 percent below 1990 levels.

The Protocol also incorporated several hastily drafted provisions to allow flexibility in how nations meet their emission limits. These included mechanisms to exchange emission-reduction obligations between nations (allowing one nation

---

[13] "Conventions" and "Protocols" are both treaties. A Convention is typically a broad agreement that provides a framework for more specific agreements negotiated in Protocols under the Convention. In this case, parties to the FCCC negotiated the Kyoto Protocol to advance the objectives and principles laid out in the FCCC.

[14] A few smaller nations negotiated particularly advantageous commitments for themselves: New Zealand's target, like Russia's, was to hold emissions at their baseline level; Norway was allowed a 1 percent increase above their baseline; Australia an 8 percent increase; and Iceland a 10 percent increase.

to make less than its required reduction, by paying the cost of a larger cut else-where). They also included provisions for nations to meet some of their obligation by enhancing carbon uptake through planting trees or similar measures, instead of reducing emissions from energy use or industry. The details of these provisions, however, along with many other matters of how to implement the Protocol, were left to be resolved later.

Further negotiations over the three years following the Protocol's signing sought to establish more specific rules for implementing the emission commit-ments, particularly regarding how much credit nations could claim for enhancing carbon uptake and for financing emission reductions abroad under the flexibility mechanisms. These negotiations brought sharp differences between two groups of industrialized countries over how much flexibility should be granted. One group, including the USA, Russia, Japan, Canada, and several other nations, sought more liberal credit for enhancing CO_2 uptake by forests or other sinks, and more flex-ibility to substitute cuts abroad for cuts at home, while most European nations wanted to allow less flexibility on each of these points.

This conflict came to a boil and negotiations between the two groups broke down at a conference in November 2000 in The Hague. Here, despite political shifts toward a harder line in Europe and the looming uncertainty of the unre-solved US Presidential election, delegates nearly reached a compromise. But the proposed compromise was rejected at the last minute by the French and German environment ministers (both Green Party members), who judged that the weak-ening of the Kyoto commitments necessary to secure US participation was too high a price. Although the breakdown of negotiations was widely blamed on the split between these two groups, it is also possible that even if this compromise had held, agreement would still have been obstructed by several other looming conflicts, both between industrialized and developing countries and among devel-oping countries, that did not come to the top of the agenda.

While the Clinton administration was confused and inconsistent in its approach to the Kyoto Protocol – as it was toward the climate issue in gen-eral – the Bush administration's attitude to the Protocol was clear hostility. Two months after taking office, the new administration announced it would not ratify the Protocol, because there was too much scientific uncertainty about climate change and because the Protocol's emission limits would harm the US economy. Although it subsequently softened its claim that scientific uncertainty supported the withdrawal, the Bush administration has continued to hold that the Protocol is unacceptable because of its high costs to the US economy, and the absence of emission limits for developing countries. In February 2002, President Bush outlined his administration's alternative approach to the issue, which included several components: a target of reducing the "greenhouse gas intensity" of the

American economy (greenhouse-gas emissions per dollar of GDP) 18 percent by 2012;[15] increased funding for climate-change science and for specific technology initiatives to reduce emissions; tax incentives for renewable energy and high-efficiency vehicles; and several programs to promote voluntary emission-reduction activities by businesses.

Following the announced US withdrawal, other signatories continued to negotiate over the flexibility mechanisms and provisions for compliance, reaching a compromise in 2002 similar to that proposed but rejected in 2000. These agreements allowed more flexibility than European delegations were previously willing to accept, and were followed by announcements that the European Union, Japan, and somewhat later Canada, would ratify the Protocol. As these sticking points have been progressively resolved, attention has shifted to more contentious points that have not yet been explicitly engaged: the form and level of emission limits after 2012, and how developing countries will participate. No significant progress has been achieved on these matters.

Still, the fate of the Protocol remained uncertain until late 2004. To enter into force – and so become binding on those who ratified – the Protocol required ratifications by 55 countries, including nations that contributed at least 55 percent of industrialized-country emissions in the baseline year, 1990. This threshold meant that, without the United States, the treaty could enter into force only if all other major industrialized countries joined – including Russia. After several years of uncertainty about its intentions, Russia submitted its ratification in November 2004, allowing the Protocol to enter into force on February 16, 2005. But while the Protocol's legal status is now secure, its contribution to an effective long-term response to climate change remains uncertain.

1.3   Plan of the book

With this background, the remainder of the book seeks to provide a clear guide to the present climate-change debate. It provides a summary of the present state of scientific knowledge about climate change, the policy options available to respond to it, the political debate about what to do about it, and how these three areas of knowledge and debate – science, policy, and politics – interact with each

---

[15] Note that this target is measured in emissions relative to the size of the US economy, not emissions themselves. The emission level that is allowed under this target grows with the economy, so if the economy grows more than 18 percent, total emissions under the target would increase. Further, the target rate of improvement is not particularly ambitious since it is roughly equal to the reduction in greenhouse-gas intensity that was realized during the 1990s.

other. Our greatest concern is with how scientific knowledge and uncertainty are used in the policy debate.

The plan of the book is as follows. Chapter 2 discusses the general characteristics of scientific debate and political debate, the differences between them, and the predictable difficulties that arise when important questions lie on the boundary between these two very different domains of argument and decision-making. Chapter 3 summarizes the present state of scientific knowledge and uncertainty about global climate change, focusing on the points that have become the most prominent matters of public controversy. Chapter 4 summarizes present knowledge and judgment about potential responses to the issue, both in the form of technological directions we might pursue and policies we might adopt. Finally, Chapter 5 does two things. First, it provides further detail about the present political debate about climate change and the foundations of the present deadlock on the issue. Second, in Chapter 5 we step back from the stance of objective reporting that we have attempted to sustain up to that point, and state explicitly our judgments of what should be done to respond appropriately to the grave threat posed by global climate change.

2

Science, politics, and science in politics

The climate-change debate, like all policy debates, is ultimately an argument over action. How shall we respond to the risks posed by climate change? Does the climate-change issue call for action, and if so, what type of action, and how much effort – and money – shall we expend? Listen to the debate over climate change and you will hear people making many different kinds of arguments – about whether and how the climate is changing, whether human activities are affecting the climate, how the climate might change in the future, what the effects of the changes will be and whether they matter, and the feasibility, advantages, and disadvantages of various responses. Although these arguments are distinct, when advanced in policy debate they all serve to build a case for what we should or should not do. Their goal is to convince others to support a particular course of action.

This chapter lays the foundation for understanding these arguments. Section 2.1 lays out the differences between the two kinds of claims advanced in policy debates, positive and normative claims. Sections 2.2 and 2.3 then discuss how science examines and tests positive claims, and how participants in policy debates use both positive and normative claims to build arguments for – and against – proposed courses of action. Section 2.4 examines what happens when these two kinds of debates overlap, as they do whenever positive claims that scientists have examined are relevant to public action – as is clearly the case in the climate-change debate. Finally, Section 2.5 discusses the role of scientific assessment in managing the boundary between scientific and policy debate. Later chapters discuss the specific claims people advance about the science and policy of climate change, and the state of present knowledge on these claims.

2.1   Justifications for action: positive statements and normative statements

The arguments that people advance to support or oppose a proposed action rest on two kinds of support: statements about what we know, or *positive claims*, and statements about what we value or should value, or *normative claims*. These two types of claim are fundamentally different. Examine the arguments advanced in any policy debate, and you will find a combination of positive and normative claims. Examine any highly contentious policy debate – like climate change – and you will find a confused intertwining of positive and normative claims.

Making a reasoned judgment of what to do about climate change requires evaluating supporting claims of both types, and recognizing the differences between the two types of claim. Although distinguishing the two types of claim can be difficult, we argue that it is essential for understanding the debate and forming an independent judgment.

A positive claim concerns the way things are: it says that something is true about the world. It might concern some state of affairs ("it is raining"), a trend over time ("winters are getting warmer"), or a causal relationship that explains why something happens ("smoking causes cancer"). Positive statements do not have to be simple or easy to verify, and they may concern human affairs as well as the biophysical world. "US foreign policy during the Cold War contributed decisively to the collapse of the Soviet Union" is also a positive statement, although one that would be hard to verify. What is essential to positive claims is that they concern how things are, not how they should be. All scientific claims and questions are positive.

A normative claim concerns how things should be: it says that something is good or bad, right or wrong, virtuous or vicious, wise or foolish, just or unjust, and so on. Examples of normative statements would include "he should have stayed to help her," "killing is wrong," "present inequity in world wealth is unjust," "we have an obligation to protect the Earth," or "environmental regulations are an unacceptable infringement on property rights and individual liberties." With few exceptions, statements or questions that include the words "should" or "ought" are normative. And the exceptions mostly involve sloppy use of language. If someone says "the Yankees should win the World Series," he probably means that they are *likely* to win (a positive claim), not that it is right or just or proper that they win (a normative claim). Of course, he might mean both these things, providing an example of how we sometimes combine – and confuse – positive and normative statements.

There are several important differences between positive claims or positive questions, and normative ones.[1] First, if a positive question is sufficiently well posed – meaning all the terms in it are defined clearly and precisely enough – it has right and wrong answers. Similarly, well-posed positive claims are either true or false. Second, the answer to a positive question, or the truth or falsity of a positive claim, does not depend on who you are: it does not depend on what you like or value, your culture, your political ideology, or your religious beliefs. Finally, arguments over positive claims can often be resolved by looking at evidence. If you and I disagree over whether it is raining, we can look outside. If we disagree over whether winters are getting warmer, we can look at the records of past and present winter temperatures. If we disagree over whether smoking causes cancer, we can look at the health records of a large group of smokers and non-smokers (who are otherwise similar), and observe whether more of the smokers get cancer.

But notice the word "often" that qualifies the above statement that positive disagreements can be resolved by looking at evidence. Looking at evidence cannot always resolve positive disagreements for two reasons, one philosophical and one practical. Philosophically, there is no rock-solid foundation for authoritatively resolving even positive questions, because you and I might disagree over what the evidence means. We might disagree over the validity of the methods used to compare winter temperatures in different places or over time. We might even disagree over whether what is happening outside right now counts as "rain." (Does a faint drizzle count? A thick fog?) If we are stuck in disagreement over such questions of evidence, neither of us can authoritatively win the argument. The best I can do is resort to secondary arguments, like what it is reasonable to believe, or whose judgment to trust, which you might also refuse to accept.

The second, practical limitation is that the evidence we need to resolve a disagreement might sometimes be unavailable, or even unobtainable. We cannot tell whether winters are getting warmer unless we have appropriate temperature records over the region and the time period we are concerned with. But while these limitations are real, they do not negate the broad generalization: looking at evidence provides a powerful and frequently effective means of resolving disagreements over positive claims.

This is not so for normative claims. Because normative questions always involve value judgments, the basis for believing that they have right and wrong answers is much weaker than for positive questions. Specific normative claims need to be

---

[1] It should be noted that all positive and normative claims can also be cast in the form of a question, for example "murder is wrong" vs "is murder wrong?" The properties of positive and normative claims are exactly the same when cast as questions. Because of that, we will talk about claims and questions interchangeably in this chapter.

based on some underlying set of principles that define the values at issue. These might be a set of religious beliefs or a moral philosophy, or might simply refer to people's preferences or interests (what people want, or what is good for them). But because people have deep differences over such underlying principles, the answer to a normative question can differ, depending on the moral or religious beliefs, the political ideology or culture, or the desires, of the person answering. Even a claim like "killing is wrong," which might initially appear non-controversial, elicits sharply differing views when considered in the context of capital punishment or euthanasia. Moreover, looking at evidence is of no help in resolving differences over purely normative questions. Normative questions are consequently more deeply contested than positive ones, and less amenable to mutually agreed resolution.

In policy debates, the arguments for and against particular actions nearly always depend on both positive and normative claims. This is because most policy choices are made for instrumental reasons: we advocate doing something because we think it is likely to bring about good consequences. Arguments about actions (Shall we raise the tax on cigarettes?) then depend partly on positive arguments about what their consequences will be (If we raise the tax, how much less will people smoke? How much revenue will be raised, from whom? How much cigarette smuggling will there be?). They also depend on normative arguments about how good or bad these consequences are (Is it fair to raise tax revenues from the poor? Is it worth accepting the projected increase in crime to gain the projected health benefits?); and on normative arguments about the acceptability or legitimacy of the action itself (Is trying to make people reduce unhealthy behavior the proper business of the government?). Similarly, people in favor of capital punishment argue that it deters people from committing heinous crimes (positive), that its application is not racially biased (positive), that procedural safeguards can reduce the risk of executing the innocent to nearly zero (positive), that murderers deserve to die (normative), and that it is just and legitimate for the state to execute them (normative). Opponents argue that deterrence is ineffective (positive), that sentencing outcomes are racially biased (positive), that the rate of errors – executing innocent people – is and will remain high (positive), and that it is wrong for the state to kill (normative).

On the climate-change issue, arguments on all sides of the debate also combine positive and normative claims. Proponents of action to reduce greenhouse-gas emissions argue that the climate has warmed, that human actions are largely responsible for recent warming, and that changes are likely to continue and accelerate – all positive claims. They also argue that the resultant impacts on resources, ecosystems, and society are likely to be unacceptably severe, and that we can limit future climate change at acceptable cost – statements that combine positive claims about the character of expected impacts and the efficacy of responses, with normative claims about the acceptability of these costs. All these claims, positive and normative, have been disputed by opponents of action to reduce emissions.

But while policy arguments may involve both positive and normative claims, these do not come neatly identified and separately packaged. Rather, many arguments intertwine positive and normative elements. For example, consider the statement, "the science of global climate change is too uncertain to justify costly restrictions on our economic growth." This says that restrictions on emissions are not justified, which appears to be a normative claim. But the claim also depends on unstated assumptions about positive matters, including what we know (and how confidently we know it) about how fast the climate is likely to change, what the impacts will be, what means are available to slow the changes, and how costly and difficult these are likely to be. The person making this argument may have considered all these things in reaching her judgment that restrictions on emissions are not justified. But hearing this argument, you would have to consider whether she is correct in these assumptions to reach an informed view of whether or not you agree with her conclusion. You and she might agree completely on what level of scientific knowledge is sufficient to warrant action, but still disagree on the conclusion if you disagree on the state of scientific knowledge.

The unstated assumptions behind an argument can be normative as well as positive. Consider the statement, "the Kyoto Protocol would cost the US economy hundreds of billions of dollars while exempting China and India from any burdens." This says something about the costs of a particular policy, which sounds like a positive claim. But the statement also has rhetorical power, since it strongly implies that it would be wrong or even foolish for the USA to join the Kyoto Protocol. Whether the statement is correct or not as a positive matter, it gains this rhetorical force from several unstated assumptions, some positive and some normative: that this cost is too high, relative to whatever benefits the Kyoto Protocol might bring the USA; that imposing the initial burden of emission reductions on the rich industrialized countries is unfair; and that other courses of action open to the USA are better.

This tangling of positive with normative claims, and of explicit arguments with powerful unstated assumptions, obstructs reasoned deliberations on public policy. It creates confusion, exacerbates conflict, and makes it difficult for citizens to understand the argument and come to an informed view. This tangling might be inadvertent, or might be intended to sow confusion in the debate, so as to obscure areas of potential agreement. The pieces of an argument cannot always be perfectly disentangled, of course. But untangling them to the extent that is feasible, and making the major assumptions that underlie policy arguments explicit, can often reduce conflict and identify bases for agreed action among people of diverse political principles.

Separating positive from normative claims is particularly important for environmental issues because of the central role positive claims play in these debates. Participants in environmental policy debates nearly always try to ground their policy arguments on scientific claims, even though the other side is often advancing directly contradictory scientific claims. In the climate-change debate, one advocate might say, "scientific evidence shows that the Earth is warming," while another says, "there is no scientific evidence that the Earth is warming." Resolving disputes over positive claims can make a substantial contribution to reducing disagreement over what course of action to pursue.

And such resolution is often possible. Indeed, on many environmental issues, the state of relevant knowledge is much more advanced and the scientific agreement much stronger than you would think from reviewing the policy debate or reading the newspaper. This is emphatically the case for global climate change. We know more about the climate, how it is changing, and how it is likely to continue changing under continued human pressures, than a look at the policy debate would suggest. To understand why, we first explore how the social process we call "science" works. We then explore how political decision-making works, and what happens when these two very different social processes come into contact with each other.

2.2   How science works

Science is a process that advances our collective knowledge of the world by proposing and testing positive claims. Science is a **social** activity – not in the sense that a party is a social activity, something we do for the purpose of enjoying other people's company, but rather in the sense that a sports team or an orchestra is a social activity: an activity that gains power from harnessing the skills and efforts of multiple people in pursuit of a common goal. The power of the social process of science to answer positive questions and advance our knowledge of the world is unparalleled in human history.

As with a sports team or an orchestra, people get to join the community of scientists by training and practising until they demonstrate that their skills and knowledge are sufficient to contribute to the group objective. Also as with a team or orchestra, there are rules and guidelines that determine how the scientific community pursues its goal and how individual scientists contribute to the collective effort. In science, the rules and guidelines make up the scientific method – a description of what scientists do that appears in the opening pages of every elementary science textbook. Although descriptions of the scientific method differ in detail, at their core all have a three-part logical structure. First, making up proposals or guesses about how the world works – these are called hypotheses or theories. Second, reasoning about what the hypothesis implies for evidence that we should be able to observe. Third, looking at the evidence to test the hypothesis, checking whether observations appear to support or refute the hypothesis.

You can use this logical structure of inquiry to investigate any positive question, small or large: "Why do my keys keep disappearing," "Who killed Cock Robin," "How do stars form," "Are people being abducted by aliens," or "Is the Earth warming?" In established communities of scientific inquiry, there are additional constraints on the application of this method that come from the collective accepted knowledge of the field. The present state of knowledge in a field, consisting of the accumulated results of all the hypotheses, observations, and tests that have been done up to now, defines what can count as an important question and a plausible, interesting answer. A hypothesis that contradicts well-settled knowledge is regarded – reasonably – as probably wrong, and so is unlikely to attract any interest. For example, a new proposal that the Earth is flat, or that the Earth is fixed in space and the heavenly bodies all revolve around it, would attract no scientific interest.

In addition, for a hypothesis to make a contribution to a scientific field, it must be *testable*: it must imply specific predictions of things you should be able to observe if it is true. It is the specific observable implications of a hypothesis that make it vulnerable to being refuted by evidence. If you look carefully and do not see what the hypothesis says you should see (or see what the hypothesis says you should not), then you conclude the hypothesis is probably wrong. Perhaps the hypothesis can be adjusted to be consistent with the evidence, but such adding of qualifications and complexity to a hypothesis to account for contrary evidence is regarded with suspicion. If a hypothesis fails to predict significant observations beyond those to which it was fitted, it will be rejected. A hypothesis that is specific, testable, and wrong can still contribute to the scientific goal of advancing knowledge. It might, for example, help to direct efforts to more fruitful lines of inquiry or stimulate someone to generate a better hypothesis. But a hypothesis with no observable implications, or whose implications are so vague or pliable that it is impossible to say what would count as decisive opposing evidence, is of no use in scientific inquiry. This is why science has nothing to say one way or the other about questions of religious belief, such as the existence of God.

Paternity testing provides a simple illustration of how evidence is used to test a hypothesis. Before DNA testing was developed, known patterns of blood-type inheritance were often used to test who was the father of a child when this was disputed. If the mother and child have certain blood types, this limits the possible blood types of the father. For example, if the mother is type A and the child is type B, then the father must be either type B or type AB. Suppose your hypothesis is that James is the father. For this hypothesis to be true, James must have blood type B or AB. If you then observe that his blood is type A, then (except for the possibility of an error in the observations) this decisively rejects the hypothesis that James is the father. Note, however, that if you find he is type B, the hypothesis that he is the father is not rejected by the evidence, but neither is it proven to be true. The true father could be James, or could be some other man with type B or AB blood.[2] This illustrates a general characteristic of scientific inquiry, that hypotheses are rejected more decisively than they are supported. Because hypotheses are constructed to imply certain observable evidence, decisive contrary evidence usually kills the hypothesis; but sometimes supporting evidence can arise by coincidence, even if the hypothesis is wrong. This characteristic is sometimes summarized by saying that science never proves anything, because while a hypothesis that has survived enough repeated testing comes to be accepted as correct, it always remains vulnerable to being disproven by some future test.

In some fields of science, the observations used to test hypotheses are generated through experiments, by isolating the phenomenon of interest in a laboratory and actively manipulating some conditions while controlling others to generate observations that are precisely targeted on the hypothesis to be tested. You can do this if you are studying chemical reactions, or the behavior of semiconductors, or the genetics of fruit flies. But for some scientific questions, such as questions about the behavior of the Earth's atmosphere, the formation of stars, or evolution of life in the distant past, you cannot do such controlled experiments in a laboratory. It is not possible, nor would it be acceptable, to put the Earth in a laboratory and manipulate some characteristic of the atmosphere to observe the response. But it is often still possible to observe naturally occurring processes in order to piece together the evidence needed to test the hypothesis.

For example, Einstein's theory of general relativity says that gravity should bend the path of a beam of light, just as it bends the path of a ball thrown into the air. The astronomer Sir Arthur Eddington saw that this part of the theory could be tested by observing the position of a group of stars when their location, as viewed from Earth, lies very close to the edge of the Sun. If light traveling from a star to the Earth bends as it passes through the Sun's strong gravitational field, then

---

[2] Modern genetic testing is much more powerful than blood-type testing, because it observes many genetic characteristics. But like blood-type testing, its results are only decisive in *rejecting* a match: if your DNA does not match all the characteristics of the tissue sample, then the sample did not come from you. If you do match all the characteristics, then the sample probably came from you, but this is not certain. In one form of DNA paternity testing, a perfect match still leaves roughly a 0.2 percent chance – two chances in a thousand – that the father is not you, but someone else who matched all the tested characteristics.

the star's position (measured relative to other stars) should appear to be shifted from when it is observed in the night sky. The Sun is so bright, however, that the only way to observe a star's apparent location when it is near the Sun is during a solar eclipse. Eddington's group traveled to Principe, off the coast of Africa, to photograph stars during an eclipse on May 29, 1919. Comparing these photographs to photos of the same stars at night showed that the light had indeed been bent by the Sun's gravitational pull, by an amount that was close to what the theory of general relativity predicted.

The work done by individual scientists or teams is only the first step in the social process of science. Whether the work proposes a theoretical claim ("I have a new explanation for the ozone hole") or an observation ("I have a new measurement of the flow of carbon between forests and the atmosphere"), it must then be judged by the relevant scientific community. This process starts with writing up the work and results – with a description of the experimental design, the data, the calculations or other methods of analysis, ideally in enough detail that someone knowledgeable in the field could reproduce the work – and submitting it for publication in a scientific journal.

The first formal control that the scientific community exercises on the quality of scientific work comes at this point. Scientific journals will not publish a paper until it has been critically scrutinized by other scientists (usually two or three) who are experts on its subject. In this process, called *peer review*, the reviewers' job is to look for any errors or weaknesses – in data used, calculations, experimental methods, or interpretation of results – that might cast doubt on the conclusions of the paper. The process is usually anonymous, so reviewers are free to give their honest professional opinion without fear of embarrassment or retribution.

Succeeding at peer review counts for everything in a scientific career. For scientific work to attract attention and respect, it has to be published in peer-reviewed journals. Proposals for research funding must also go through peer review. For scientists to get and keep jobs and achieve all other forms of professional reward and status, they must succeed at getting their work through peer review.

Aside: how tough is peer review really?

Very tough. You might think peer review is a rubber stamp, or a comfortable process by which scientists pat each other on the back. On the contrary, peer review is a careful, highly critical examination of the work being proposed for publication. The following rejection letter from a journal editor (slightly edited for clarity and anonymity) gives a taste of how demanding the process is.

Dear Dr. Smith,

   I am now in receipt of the reviews of your paper entitled, "Isotopes, seasonal signals, and transport near the tropical tropopause". On the basis of these reviews I regret that I cannot accept this paper for publication in the Journal of the Atmospheric Sciences in its present form. This was a very difficult decision, since Reviewers A and B recommend rejection, while Reviewer C is much more positive about the study. Yet even Reviewer C has serious misgivings about the potential for numerical problems in the model, and cites insufficient comparison and justification of the results with respect to observations. For their part, Reviewers A and B are thoroughly unconvinced that the model is sufficiently constrained by the limited observations available. Furthermore, the reviewers are concerned that the model's extreme sensitivity to many tunable parameters renders the results highly suspect. Given the seriousness of these issues, I cannot accept this manuscript. However, since Reviewer A has suggested that the study could be reworked to something acceptable and Reviewer C is generally supportive, I encourage you thoroughly to revise the paper and resubmit a new version – if, that is, you think the concerns can be adequately dealt with. In that regard, Reviewer A argues for a much more complete sensitivity analysis, and all the reviewers call for detailed justification of the many decisions made in tuning the model. This should be done with reference to observations as much as possible, but barring that possibility, physical arguments and results from previous studies could also be used. If you choose this course, I suggest that you pay careful attention to all the major and minor comments of the reviewers. You should also provide a detailed, point-by-point response to each reviewer.

Regards,
John Q. Pseudonym, Editor

What does this mean? Reviewers A and B were not convinced that the scientific analysis supported the conclusions. Although reviewer C recommended that the paper be accepted, the editor looked carefully at the reviews and the paper, decided he agreed with reviewers A and B, and rejected the paper. But while this version is not acceptable, the authors might still succeed at making the work publishable. The editor advises them to revise the work, address the reviewers' criticisms, and try again.

Peer review is a highly effective filter, which stops most errors from being published, but it cannot catch every problem. Reviewers occasionally fail to notice an obvious mistake, and there are some types of error that reviewers usually cannot catch. They cannot tell if the author misread observations of an instrument, or wrote a number down wrong, or if chemical samples used in an experiment were contaminated. Moreover, peer review often cannot identify clever fraud, such as the rare cases where the scientific work being reported was not really done at all.

But peer review is only the first of many levels of testing and quality control applied to scientific claims. When an important or novel claim is published in a journal, other scientists test the result by trying to replicate it, often using different data sets, experimental designs, or analytic techniques. While one scientist might make a mistake, do a sloppy experiment, or misinterpret their results (and peer reviewers might fail to catch it), it is unlikely that several independent groups will make the same mistake. Consequently, as other scientists repeat an observation, or examine a question using different approaches and get the same answer, the community increasingly comes to accept the claim as correct.

For example, during the early years of controversy over ozone depletion in the 1970s, the available observations showed no decrease in global ozone had occurred. Although the theory suggested that continued releases of chlorofluorocarbons (CFCs) would lead to a reduction in ozone, no reduction could be seen at that time. In the early 1980s, a few scientists began proposing that a decline could be observed in the latest ozone measurements. There were many problems with the data, however, and when other scientists examined the data, they concluded that the reductions being proposed were not well founded. As a result, the claims were rejected. Then in 1988, a new analysis including more recent data suggested stronger evidence of a decline. Because this claim was so important, three other scientific teams checked and re-analyzed the data behind this new claim, as well as analyzing related data. This time the other teams also found a decrease in ozone, similar in size to that calculated by the first team. The conclusion was therefore confirmed, and atmospheric scientists accepted that there now was a real decline in global ozone.

This multi-layered process of criticizing, testing, and replicating new scientific claims is **public, collective, and impersonal**. Individual scientists make mistakes, and are prone to biases, enthusiasms, or ambitions that may cloud their vision, as we all are. But however intensely a scientist may hope for honor from having his novel claim accepted, or want a result consistent with his political beliefs or financial interests, scientists know that any claim they propose, especially if it is an important one, will be critically examined by other scientists and sloppy, biased, or weakly supported work is likely to be exposed. Moreover, scientists confer respect and status on their peers who are careful in their work, critical and fair in their argument, and cautious in advancing claims. Intemperate claims, partisan or biased testing, or less than scrupulously honest reporting of results can so severely damage a reputation that scientists have strong incentives to be cautious. The result of this process of collective testing, and the incentives embedded in it, is to make science highly conservative. The burden of proof lies with the person making any claim that extends present knowledge or contradicts present belief. The more important and novel the claim being advanced, the more aggressive the scrutiny and testing it will receive and the higher the standard of evidence required to accept it: remarkable claims require remarkable evidence. This is the way science maintains the stability of the received body of knowledge and protects against errors and fads.

Aside: is this really how science works?

Not exactly, but close enough. This description is a simplification – some would call it a dogma – of how science is actually practised. Several decades of research in the sociology of science has fleshed out how and how much the actual practice of science diverges from this model, in particular how social factors impinge on the practice of science. The most basic insight was that of Thomas Kuhn, who recognized that normal progress in a scientific field depends upon a deep level of shared assumptions that define what questions are important, what lines of inquiry are promising, and what hypotheses are plausible and interesting. These shared presumptions, which Kuhn called "paradigms," are not explicitly examined or even necessarily recognized by the scientists who hold them. Paradigms change only infrequently, in revolutionary periods that follow the accumulation of some critical mass of "anomalies" – results that don't fit the accepted model, but are provisionally set aside.

Science is not an abstract, rational process, but a collective human endeavor. As such, social factors such as status, charisma, and rhetorical skill influence to some extent whose arguments get paid attention and trusted; not all claims are immediately tested and replicated; and consensus views of what questions are interesting and important are not formed on purely rational bases. But the power of these social processes to influence the content of what comes to be accepted as scientific knowledge is limited and provisional. Conspicuous claims that do not stand up to testing eventually get rejected, no matter who is supporting them. Accepted beliefs that accumulate enough anomalies are eventually re-examined, revised, or rejected, however comfortable or fashionable they may be.

This process of making hypotheses and using disciplined, repeatable, observations to test them does not generate proven truth. Even a hypothesis that has survived repeated testing and come to be accepted as correct remains vulnerable to being overturned by some future test. But some claims are so well tested and verified, by accumulating independent evidence, resolving controversies, and rejecting contrary claims, that they come to be regarded as facts. For example, we now accept as facts that the structure of DNA is a double-helix, that atoms obey the laws of quantum mechanics, and that humans activities have significantly increased the abundance of CO_2 in the atmosphere. These claims have been so well verified that further testing of them is considered unnecessary and uninteresting.

How can you tell whether the relevant scientific community has accepted a claim as "true"? The most reliable way is to look in the peer-reviewed literature for multiple, independent, peer-reviewed verifications. Other factors count in addition to the number of verifications, of course. Some tests are more stringent than others, for example, as DNA paternity testing is more stringent – more likely to reject a match – than blood-type testing. When newer, well checked observations contradict older ones, the newer ones are usually given more weight because observational instruments generally improve as technology advances. The extent to which competing claims have been tested and rejected also matters – if every proposal but one has been rejected, the remaining one is more likely to be accepted, at least tentatively, even if the affirmative evidence supporting it is far from conclusive.

The reputations of the scientists involved also affect the community's willingness to accept a claim. The same work is likely to be given more credence when done by a scientist with a well-established reputation for careful, competent work than when done by one who is unknown or known to have done sloppy work in the past. Eddington's confirmation of general relativity carried more weight, and was probably accepted more quickly, because of his preeminent reputation for scrupulously careful observations.

Even though scientific knowledge is always provisional, never truly proven, a strong scientific consensus provides a better basis for relying on the truth of a positive claim than is provided by any other human process for pursuing knowledge. It is of course possible for a strong scientific consensus to be wrong. We are reminded of this possibility on those occasions when later results contradict and eventually overturn a previously accepted understanding. But while this does happen, and generates much excitement and attention when it does, it occurs infrequently.

The risk of a consensus later being found to be wrong is greater for some types of scientific claims than for others. The risk is greatest for fundamental theories, particularly if their strongest predictions concern matters beyond our present capability to observe. We would not be too surprised if Einstein's theory of gravitation were one day superseded by another, just as Newton's theory of gravitation was previously superseded by Einstein's. The risk is smallest for simple, concrete claims, such as a single observation or measurement, such as a measurement of the charge of an electron. When an observation has been repeatedly checked using various methods and accepted as correct, it is quite unlikely to be subsequently overturned. Between these extremes, claims about the associations between different observations from which we infer cause and effect are on somewhat weaker, although still very strong, ground. We would be extremely surprised to learn in the future that smoking does not increase the risk of cancer, or that chlorine chemistry does not cause the Antarctic ozone hole, although it could happen.

The lesson we draw from this discussion is that, when there is a strong scientific consensus on some positive point, those outside the relevant scientific community should rely on it. This advice applies in particular to policy actors dealing with contested decisions that rely in part on positive points of scientific knowledge, such as the climate-change debate.

Unfortunately, this advice is not always helpful. There are two principal reasons: a strong consensus on a policy-relevant scientific question might not exist; or a consensus might exist but be difficult for those outside the field to observe. A consensus might not exist because a key question lies beyond present scientific knowledge or research capabilities, or because it has simply not attracted much scientific effort. It is not always the case that the positive questions of greatest importance to policy decisions are also of high scientific interest. Alternatively, a question might be under investigation, accumulating evidence but not fully resolved, perhaps with scientists disagreeing over how well settled it is. One implication of the cautious and conservative nature of science is that acceptance of new claims happens slowly, often much more slowly than answers are demanded by the policy-making process.

Alternatively, a consensus might be present in a scientific community but difficult for anyone outside the field to observe. Scientific discussions focus on what is interesting to scientists: not what is well established, but what is new, uncertain, and controversial. Moreover, even a strong consensus on some point may be obscured for those outside the field by a few vocal advocates of an opposing view, even if that opposing view has been examined and decisively rejected. Since even scientists in other fields may lack the specific knowledge to review and understand the peer-reviewed literature to judge the merits of opposing claims, non-scientific policy actors cannot hope to make such independent judgments themselves. Rather, they must rely on some form of summary and synthesis of what the scientific community knows and how confidently it knows it. Sections 2.4

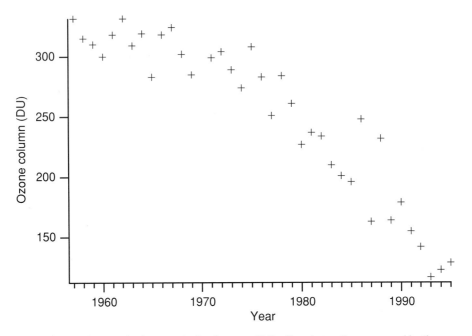

Figure 2.1. Total atmospheric ozone in October over Halley Bay, Antarctica, measured by the British Antarctic Survey. The amount of column ozone is expressed in Dobson Units (DU). Source: adapted from Fig. 7.2 of Dessler (2000).

and 2.5 provide further discussion of this problem, and the role of scientific assessment bodies in resolving it.

An example: the discovery and explanation of the Antarctic ozone hole

The discovery and verification of the Antarctic ozone hole and the search for its explanation illustrate the checking of scientific claims and the logic of testing scientific hypotheses.

In 1982, researchers at the Halley Bay station of the British Antarctic Survey noticed that the total amount of ozone over their station in October – early spring in Antarctica – appeared to be dropping sharply from levels of the 1960s and 1970s. (Figure 2.1 shows their data, extended to the mid-1990s.) Ozone in other months of the year appeared to be normal. Initially concerned that they might be seeing an instrument error – always a likely explanation for any wildly unexpected scientific observation – the scientists spent two years checking, confirming, and extending their results before submitting them to the journal *Nature*, where their paper was published in June 1985.

Their observations were wholly unexpected, and created a scientific firestorm. The results had passed peer review, but it was still imperative, given the dramatic nature of the claim, to obtain independent verification of the losses. This was quickly obtained by review of archived data records from a satellite instrument measuring ozone. Several other groups made additional measurements confirming the losses over the next few years.

So the losses were real, but what was causing them? Atmospheric scientists had been predicting for ten years that we should be starting to see ozone depletion from chlorofluorocarbons (CFCs). But these observed losses, in addition to being much more extreme than what was predicted, were also occurring in a location and at a time that was virtually the opposite of where the theory of CFC-induced ozone loss predicted them. Consequently, it was not clear whether CFCs had anything to do with the observed losses.

Over the following year, three competing theories were advanced to account for the shocking losses, each implying specific things that should be observable in the region of ozone loss. Observations made in the Antarctic in 1986 and 1987 decisively rejected two of the theories, and provided compelling support for the third.

The first theory proposed that the ozone was being destroyed by naturally occurring nitrogen oxides, whose concentration in the Antarctic stratosphere increased following the peak periods of solar radiation that occurred every 11 years. If this theory was true, several pieces of evidence should have been present. Ozone loss should be greatest at the top of the stratosphere, because this is where most nitrogen oxide production occurs, and should be accompanied by elevated concentrations of nitrogen oxides. Moreover, there should have been similar ozone losses following previous solar-maximum years such as 1958 and 1969, and losses should have begun to reverse by the mid-1980s. Observations in 1986 quickly ruled out this hypothesis, since none of these predictions was found to be true: ozone losses were most extreme in the lower stratosphere and were accompanied by low, not high, concentrations of nitrogen oxides; moreover, review of historical ozone records showed no sign of the predicted earlier losses, and losses were clearly accelerating through the mid-1980s.

The second theory proposed that changes in the circulation of air through the global stratosphere were reducing the transport of ozone to Antarctica. This theory also predicted several things that should be observable if it was true. First, temperatures in the depletion region should be unusually cold. Second, the vortex of stratospheric winds that surrounds Antarctica in winter should be stronger than usual and persist longer into the spring. And most crucially, the general movement of air under and within the region of depletion should be upward. Although early observations gave limited support to this theory – the vortex did appear to be unusually strong, and temperatures in October appeared to have grown colder (although not in August or September) – this theory was rejected by 1987 when it became clear that Antarctic air was generally sinking, not rising as the theory required.

The third theory proposed that ozone was being destroyed by newly identified chemical reactions in which chlorine carried to the stratosphere by CFCs acted as a catalyst. Although several groups of scientists proposed different specific chemical reactions, these all required that a particular chemically active chlorine species, chlorine monoxide (ClO), should be abundant where the ozone loss was occurring. Simultaneous observations of ozone and ClO from flights through the ozone hole in September 1987 found remarkably strong support for this hypothesis. ClO suddenly increased a hundred-fold each time the aircraft crossed into the region of ozone loss, and dropped again as it left the region. This negative correlation between ClO and ozone (whenever ClO went up, ozone went down) was so exact that many observers called this result the "smoking gun" – one of those rare instances in which a hypothesis comes to be accepted as true on the basis of a single, compelling observation. Over the following few months, as the results were reviewed for publication and intensively discussed at several scientific meetings, there formed a strong consensus, which since that time has been sustained and further elaborated in its details, that CFCs and related chlorinated chemicals were the principal cause of the ozone hole.

2.3   Politics and policy debates

Politics is concerned not with positive questions, but with collective action: not what is true, but what shall we do? Politics embraces the processes of argument, negotiation, and struggle over joint actions or decisions – most often the decisions of what policies will be adopted by government institutions. Like most competitive arenas, politics involves conflict but the conflict is bounded. In competitive sports, the boundaries are defined by the rules of the game as enforced by the officials. In politics, the boundaries are defined by the structure of rules and institutions within which decisions are made. For example, national political decision-making takes place within a complex structure of constitutions, laws, and traditions that grants specific authorities to make decisions and imposes various constraints on the exercise of that authority. Sometimes the authority to make decisions is simple and absolute: the President of the United States has the power to grant pardons for certain criminal offences. More often, however, authority is limited by various rules or conditions that restrict its exercise, or by related authority held by others. Under the US Clean Air Act, for example, the Environmental Protection Agency (EPA) has the authority to enact regulations to restrict chemicals that deplete the ozone layer. But to restrict a chemical, the EPA must present evidence that the chemical is a strong enough ozone depleter to fall under the requirements of the law, must publicize the proposed regulations for a 90-day comment period, and must respond to the comments received before enacting the final regulation. If the EPA fails to meet these requirements, the regulation might be overturned in court. More broadly, since the EPA's authority to make these regulations is delegated to them by Congress in the Clean Air Act, Congress could revise or revoke the authority by amending the Act.[3]

Sometimes, particularly for issues on which the government has not previously acted, the relevant authority might be widely distributed, defined only vaguely, or not defined at all. The more novel the issue and the less clear the existing lines of authority, the more fluid is the political process in terms of what the key decisions are, who makes them, who gets to influence them, and what factors contribute to the outcome. Political systems differ in their overall openness to public influence over decisions, and in their particular channels for influence. Even in highly open systems such as the United States, however, exercising influence over a policy decision is extremely difficult, and requires a great deal of time, energy, money, strategic skill, or luck. Because it is so difficult, those who do mobilize to influence policy on any particular issue represent a tiny fraction of the electorate.[4]

The few people who do mobilize to influence policy decisions do so for many reasons. Some may hold strong views about the right thing to do or the best interests of the nation. Some may expect proposed decisions to benefit or harm them in some concrete way – for example, affecting the health or well-being of their families, helping or hurting their livelihoods, or affecting the value of their property or the profitability of their business. Some may have ambitions to exercise political power or influence. Any of these motivations can bring conflicts between groups seeking to influence policy. My group might compete with yours to be a

---

[3] Since this section of the law meets a US obligation under an international treaty – the Montreal Protocol – changing the law to revoke this authority would be likely to put the United States in violation of the treaty. But while this obligation might make Congress think twice about making such a change, it does not eliminate their authority to do it, since Congress and the President also have the authority to withdraw from international treaties.

[4] That few people get active on an issue does not, however, mean that few care about it: public opinion comes in all shapes and sizes. No issue gets onto the political agenda without at least a few people caring strongly about it. But the rest of the public might or might not understand the issue or care about it, and those who do care might be largely in agreement or strongly divided. Those who are active on an issue try to portray their position as strongly supported by a "silent majority" when they can. Public opinions tend to be strongest and most divided on issues that raise deep differences of principle, such as abortion or capital punishment. While most people express strong support for most issues of environmental protection, this only infrequently rises to the intensity of opinion typical of more morally charged issues.

key deal-maker on environmental policy; a proposed regulation might benefit my industry but harm yours; or I might passionately believe that the environment needs further protection, while you believe with equal passion and sincerity that environmental regulations threaten basic liberties.

People sometimes speak, as we do here, about the "policy debate" on important public issues like climate change. But describing policy-making as a debate is somewhat misleading. You win a debate by persuading people – impartial third parties, and sometimes your opponents – that the arguments favoring the course of action you support are stronger than those on the other side. In policy-making, the strength of arguments on your side also clearly matters, but as one factor among many that influence what happens. Policy actors use many methods to build support for the decisions they want: well-founded rational arguments when these are available, but also biased or inaccurate arguments, anecdotes and stories, invocation of powerful symbols, appeals to emotion or prejudice, flattery and manipulation, promises and threats, deals to exchange support on other issues, and sometimes – although these are illegal in most nations – bribery and coercion.

But in this complex mix of factors shaping policy decisions, rational arguments almost always matter at least a little, and sometimes a lot. Rational arguments matter more when an issue is prominent enough, and is perceived to have high enough public stakes, that it attracts public attention and media scrutiny. Such scrutiny increases legislators' and officials' concern with acting competently and impartially in the public interest – and being seen to act so – and so reduces the scope for the more crass and sneaky forms of political influence that would be embarrassing if revealed. Rational arguments also matter more when an issue is so novel that its character, relevant analogies to other issues, and the consequences of alternative actions, are unclear. Under these conditions, many policy actors may be uncertain what choices to favor: their general political principles may give little guidance, and their interests in any particular choice might be unclear. Climate change has these characteristics: enough salience that political decision-making is subject to heightened scrutiny, enough novelty to challenge existing lines of authority, and enough uncertainty that many policy actors do not line up predictably according to either their general political principles or how the issue is going to affect them.

Consequently, climate-change is an issue on which we expect rational argu-ments, both positive and normative, to be influential. On normative arguments, an explicit debate between competing perspectives, some of which may be directly in contradiction, makes sense. If one group in the climate-change debate argues that our primary obligation is to protect the environment, while another argues for the primacy of individual freedoms, neither of these is right or wrong. It is entirely appropriate for proponents of these contending views to compete to

persuade policy-makers and citizens. Science and scientists have no special authority in this debate. On relevant positive arguments, however, such as observed
trends and variations in the climate, the causes of these trends, and the likely
nature of future climate change under additional human inputs, scientific knowledge does have special authority and therefore much to contribute. But this potential contribution is often obstructed by a lack of understanding of the differences
between scientific arguments and policy arguments.

The most important of these differences concern the motivations of participants, and the rules that constrain their conduct. Scientists gain professional
status by advancing shared knowledge, but also by being cautious in interpreting new claims and fair in their assessment of competing claims. While there are
real disagreements and rivalries in scientific debates, these norms introduce substantial elements of shared interest, so scientific disputes are rarely "zero-sum".
Advancing knowledge benefits the scientific enterprise as a whole, even if the one
making the discovery benefits the most.

In policy debates, there can also be important shared interests – for example,
in not wasting tax dollars on worthless projects or corruption, and in protecting the state and its citizens against clear threats from hostile foreign powers –
but competing interests are much more prominent. Political actors are rewarded
for succeeding in various ways where one individual's or group's gain is usually
another's loss, such as gaining and holding power, enacting policies consistent
with their political principles, and delivering the benefits of state action such
as spending on public works projects to their supporters and constituents. Even
when an issue like climate change is so novel and uncertain that people see less
clearly where their material interests lie, these incentives still introduce competitive elements into all policy decisions.

A second fundamental difference between scientific and political arguments
concerns the rules of acceptable argument. The rules in both domains are mostly
unwritten, enforced only by the approval and censure of others, but they are nevertheless real and powerful. The rules of scientific argument are highly constraining.
Whatever their true motivation, scientists must argue as if motivated purely by
the pursuit of knowledge. A scientist who breaks the rules – who makes intemperate claims based on limited evidence, fails to acknowledge how he or she could
be wrong, selects evidence opportunistically to support his or her view, ignores
or misrepresents contrary evidence, makes emotion-laden arguments, or makes
personal attacks on opponents – risks irreparable harm to his or her reputation
and professional standing.

The rules of policy argument are much more lenient. In policy debates, exaggerated, selective, or biased claims, appeals to emotion, and personal attacks of some
minimal relevance to the matter at issue, are not just frequently effective; they

are also rarely punished or even censured. Even more aggressive tactics, such as personal attacks with no substantive relevance, appeals to prejudice, and outright lies, are only weakly restrained. Public outrage might do so, but rarely lasts long enough to be effective, while political opponents must be careful in calling such tactics to account, since they may sometimes use them too. Moreover, there are many ways to gain standing in a policy debate. Having a reputation for knowledge and honesty is one way, but so is representing an important constituency, or the self-fulfilling perception that you are likely to be influential. Consequently, losing scientific credibility does not necessarily jeopardize standing in policy debates. In view of the lesser consequences from going over the line, policy advocates are much more willing than scientists to take risks with their credibility in order to advance their objectives.

2.4   When science and politics meet

When the consequences of alternative policy choices depend on positive matters that are areas of active scientific research, policy-making requires some interchange between scientific and policy debate. While many areas of policy decisions require such interchange, it is essential in deciding how to respond to environmental risks like climate change, since the desirability of different courses of action depend on how much and how fast the climate is likely to change, the nature and severity of the resultant impacts, and the availability, effectiveness and costs of policies we might adopt to slow the change and adapt to its impacts.

This need for interchange between scientific and political debate poses challenges to both science and politics, due to the different goals and rules of the two domains. These challenges go in both directions. Politics poses hard challenges to science, because the questions most relevant to policy decisions may not have scientifically well-founded answers. Sometimes such questions are so far beyond present scientific capabilities that solid answers are unlikely for years or decades, perhaps ever. Confident, precise predictions of future fine-scale climate change probably fall in this category: how will the climate change where I live, and what will the consequences be? But political debate demands fast answers and is unsympathetic to scientific caution. Because many scientists prefer to focus on their scientific work and view the political arena with some mixture of fear and contempt, the few who choose to comment publicly represent an odd mixture of motivations. Some are civic-minded and courageous; some believe their scientific status gives special weight to their policy views; and some seek the public forum for its own rewards – the fame or notoriety, sometimes the influence or profit. Moreover, in policy arenas it is difficult to distinguish well established scientific claims from those that are unrepresentative, uninformed, eccentric, or outright dishonest. Even when scientific consensus on a point is powerful and well founded, it can be difficult to persuade a lay audience when arguing against a rhetorically skilled opponent, particularly one who operates by the more lenient rules of policy debate.

Individual scientists who wish to contribute responsibly to public debate, or are called on to do so, consequently face a nasty bind. They can refuse, and shun what is arguably their civic responsibility. They can try to reflect the state of knowledge and uncertainty responsibly, and risk having their scientific caution be taken as indecisiveness. Or they can set aside their scientific conservatism and try to distill their understanding into simpler terms more likely to resonate and be understood in the policy domain, and risk being charged as intemperate or fame-seeking. Some scientists try to resolve this dilemma by drawing clear distinctions between their roles as scientist and as citizen, expressing carefully qualified scientific opinions then explicitly changing their stance to speak more simply and forcefully as a concerned citizen, but this distinction is extremely difficult to draw cleanly. A speaker's scientific standing unavoidably provides some additional credence to their policy views, even while using scientific standing to gain a platform in this way puts that standing at some risk. Moreover, those who try to draw this distinction between scientific and personal views are still frequently censured for politicizing their science, sometimes by their scientific colleagues and more frequently by policy actors with opposing policy views.

In addition to the challenges that policy debates pose to science, science also poses hard challenges to policy debates, because citizens and politicians are not generally able to make independent judgments of the merits of scientific claims. With rare exceptions, policy actors do not have the time or training to read the peer-reviewed literature and evaluate the contending claims in it. Consequently, any attempts they make to independently evaluate scientific claims carry a large risk of error, because even completely spurious claims can seem plausible to someone who does not know the field. Once again, the history of the ozone-layer debate provides a vivid example. In 1974, within months of the first suggestion that chlorine from chlorofluorocarbons (CFCs) could destroy stratospheric ozone, political opponents of CFC restrictions began circulating the opposing claim that this was impossible because CFC molecules are so much heavier than air that they could never rise to the stratosphere. This sounds like common sense: nearly everyone has seen a mixture of fluids of different densities, like oil and vinegar in salad dressing, in which the heavier ones settle to the bottom. But common-sense or not, this claim is obviously false to anyone who knows a little about how the atmosphere behaves. The atmosphere is not a quiet isolated vessel, but is continuously mixed by winds, vertically as well as horizontally, so gases are mixed to uniform concentrations from the surface to well above the stratosphere, regardless of their weight. Despite its obvious falsehood to anyone with the relevant knowledge, this claim continued to circulate for more than 15 years as supposed evidence that CFCs could not harm the ozone layer, and still reappears now and then.

But while policy actors usually cannot evaluate scientific claims, they nevertheless actively seek scientific support for their positions, especially when the stakes are high. This can be a powerful rhetorical device, since any position that can take on science's reputation for disinterested pursuit of truth will appear more persuasive. But the lenient standards of argument in policy debates give powerful incentives to advance spurious or eccentric scientific claims if strong ones are not available, and provide ample opportunity to attack and seem to discredit opposing claims even when those have strong scientific support. With one claim seeming to cancel an opposing claim, even a settled argument can be made to look like a draw.

The status quo enjoys a large advantage in any policy debate, in that it takes substantial political energy to make any change. Consequently, an advocate of the status quo may succeed by simply persuading people that the scientific evidence is too "uncertain" to justify a change. This can sometimes be achieved merely by advancing enough arguments, even if they are all bad ones, to confuse people. In seeking support for their positions, policy actors consequently have a great deal of liberty. They can selectively scan the hundreds or thousands of scientific papers published on a particular controversy to find a few that support their case, even if these are old, known to be erroneous, or decisively refuted by other work. Of the thousands of people with scientific credentials, they can usually find a few who are contrarian or opportunistic enough to go on the record making claims that virtually everyone working in the field knows to be false. As recently as a few years ago, there were still a few scientists willing to state publicly that there was no scientifically persuasive evidence of a link between smoking and cancer. Advocates with enough resources can finance such individuals' participation in policy debates, or even fund programs of research they think likely to generate favorable results. As a consequence, many contending scientific claims circulate in policy debates. Their merit might be highly variable, but the policy arena provides no way to evaluate them and reject even those that an overwhelming scientific consensus has judged to be false.

In view of the widespread use of false or misleading scientific claims in policy debates, a policy actor who wants to develop an informed view of the state of scientific knowledge has no choice but to rely on some level of trust, either in individuals or in institutions. But it is hard to know whom or what to trust, and how much to trust them. Facing a cacophony of contending claims, policy actors' options are limited and unattractive. Some may have a trusted advisor with relevant scientific expertise, but on any given issue or controversy most will not. Some may simply accept the claims that are consistent with their policy preferences, or side with others who share their political views – but this approach is dangerous, since there is no basis for confidence that that actual state of scientific knowledge will match their political views. Some may simply withdraw from the issue and leave it to others to fight it out.

The press is often of little help. Journalists frequently do not understand scientific issues any better than policy actors. Even when they do, journalists follow a professional norm of providing balance between opposing views. Moreover, controversy sells newspapers. Since even settled issues may be debated by a minority, the press generally underreports scientific consensus. Worse still, coverage often favors the dramatic, so the press may give particular prominence not just to minority views, but to extreme views. One scientist's speculation that global climate change may trigger a sudden return to ice-age conditions, or the presentation of such an unfounded scenario in a popular film, makes for dramatic coverage. So do the claims of a half-dozen "climate skeptics" that the scientific consensus on climate change is a political conspiracy, or the repetition of these claims in a best-selling novel. The careful reporting of the content of that consensus and the evidence supporting it do not.

The unfortunate result of this rough meeting of science and politics is an exaggerated and misleading appearance of scientific controversy and conflict played out in policy debates and in the press. Claims and counter-claims are presented without regard for the strength of their evidence or the numbers and stature of the scientists supporting them. Well-established claims backed by near-universal scientific consensus cannot readily be distinguished from the views of tiny partisan minorities. Many citizens and policy-makers consequently perceive rampant ignorance and uncertainty even where much is well known, and perceive serious disagreement even where there is overwhelming consensus. With the degree of real scientific knowledge and agreement not recognized, the potential for scientific knowledge to illuminate policy debate and delimit conflict is frequently not realized.

2.5   Limiting the damage: the role of scientific assessments

The picture we have painted is bleak, but far from hopeless. Our view, which provides the motivation for writing this book, is that the climate-change issue is more confused and contentious than it needs to be, because of widespread misrepresentations of the state of scientific knowledge on relevant positive questions. Because of the different rules of scientific and policy debates, policy actors have ample opportunity and incentive to misrepresent scientific knowledge, by advancing claims that have been clearly rejected, misrepresenting uncertainties (in some cases exaggerating them, in others minimizing or ignoring them), or exaggerating scientific dissent when the implications of the consensus are unfavorable for their preferred policies.

Such tactics can be effective, particularly when the implications of the arguments align with the audience's broad political principles and prior beliefs. Those whose broad political views are to support free enterprise and oppose government regulation tend to be sympathetic to claims that the evidence for climate change is weak and any future changes are likely to be small and manageable. Those whose broad politics favors stronger government regulation of business and industry tend to be sympathetic to claims that climate change is happening and is likely to be severe. With these tactics widespread, policy actors who do not want simply to act on the basis of general political principles, but to consider the actual state of present scientific knowledge, can have difficulty learning what that is. Deciding what to do about global climate change would be difficult and contentious enough even without these misleading tactics, but these tactics make it worse.

It is both worthwhile and feasible, however, to structure policy debates so as to reduce the incentives and opportunities of policy actors to practise such deceptions. One essential key to such improvement lies in disentangling the policy debate into separate, precisely posed questions and noting which of these are positive questions of scientific knowledge about the world, and which are normative questions of our values, desires, and political principles. It is not always possible to draw these distinctions perfectly cleanly, but trying to do so to the extent feasible can bring large benefits. For individuals engaged in the policy debate, attempting such separation of positive and normative claims will help to understand arguments that others are advancing, and provide a better basis for deciding whom to trust, to what degree, on what questions, and coming to an informed view of what decisions to favor. And for policy debate overall, pursuing such separation of questions is likely to reduce confusion and conflict, and provide a sounder basis for seeking courses of action that might gain broad support.

To the extent that some distinction between positive and normative questions can be maintained, the two types of question are appropriately dealt with in different ways. Positive questions – such as the evidence for present climate change, the changes that are likely over coming decades, and their consequences – are best examined by scientific processes, not democratic ones. For such questions, when a strong consensus exists among the relevant scientific experts, this is the closest thing we have to well-founded knowledge, and it is entitled to substantial deference in policy debates. Strong expert consensus does not, of course, always exist on all policy-relevant positive questions. When it does not, the best indicator of the state of scientific knowledge is the range and distribution of judgment among relevant experts, to the extent that this can be specified. When policy decisions have high stakes that depend on the answer to a positive question – for example, how will the climate change under various alternative emission futures – policy actors can do no better than to take such a consensus or distribution of expert judgment as a correct representation of present knowledge and uncertainty.

The problem with this advice is that policy actors cannot reliably observe the state of scientific consensus or the distribution of expert judgment on specific policy-relevant questions. It is simply too difficult and time-consuming for those without specialist training to digest the peer-reviewed literature directly to judge the state of relevant scientific debates. Rather, policy actors must rely on some type of scientific advisor or scientific advisory process to tell them. The process of synthesizing, evaluating, and communicating scientific knowledge to inform a policy or decision process is called *scientific assessment*.

Scientific assessments connect the domains of science and democratic politics, but are distinct from both. They differ from science because rather than advancing the active, contested margin of knowledge on questions that are important for their intrinsic intellectual interest, they seek to make consensus statements of present knowledge and uncertainty on questions that are important because of their implications for decisions. They differ from democratic policy debate because they reflect deliberation over positive questions among scientific experts based on their specialized knowledge, not among all citizens or their representatives over what is to be done.

The need for effective scientific assessment to support environmental policy-making at both the national and international level has been widely recognized for at least 25 years. There are many ways to conduct scientific assessments, and many bodies that do them. The US federal government often calls on the National Academy of Sciences to assemble expert panels to provide advice on scientific and technical matters related to national policy, and has sometimes established special scientific assessment bodies on particularly important or contentious issues. Scientific assessments can play an even stronger role in international environmental policy-making, because they can make authoritative statements of scientific knowledge that transcend differences in national policy positions. Atmospheric-science assessment panels on stratospheric ozone played a particularly influential role in the establishment and subsequent revision of international agreements to control ozone-depleting chemicals, by making highly visible, authoritative statements of how certain key scientific points, which were previously contested in the policy debate, had been resolved.

Assessments do not always succeed at making effective contributions to policy debates. They can fail to do so in many ways. For example, some assessments lose credibility by making explicit policy recommendations or otherwise going beyond their authoritative expertise. In contrast, others fail to synthesize present knowledge into a coherent summary view, often out of reluctance to state explicit judgments of the relative strength of present contending claims or the distribution of present knowledge. When assessments succeed, they do so by effectively managing the boundary between scientific and political debate. There is no single model of how to achieve this, but several areas of skill and judgment can make strong contributions to success. Leaders of successful assessments must maintain an alert ear to identify positive questions that policy actors perceive to be of high relevance. They must be able to motivate scientific participants to a level of synthesis and integration that is rarely done explicitly in purely scientific forums, while still maintaining rigorous standards of scientific debate. They must also exercise effective judgment to stay within their domain of expertise, and must be able to communicate clearly to a non-scientific policy audience without sacrificing scientific accuracy.

The Intergovernmental Panel on Climate Change or IPCC, whose establishment we discussed in Chapter 1, is the primary body responsible for international scientific assessments of climate change. Since its establishment in 1989, the IPCC has undertaken three full-scale assessments of climate change – in 1990, 1995, and 2001 – as well as many smaller and more specialized reports. Each of the full assessments is a huge undertaking. The reports involve hundreds of scientists from dozens of countries as authors and peer reviewers, including many of the most respected figures in the field. These groups work over several years to produce each full assessment, and their reports are subjected to an exhaustive, publicly documented, multi-stage review process. In view of the number and eminence of the participating scientists and the rigor of their review process, the IPCC assessments are widely regarded as the authoritative statements of scientific knowledge on climate change. We will refer to these assessments repeatedly in summarizing present scientific knowledge on specific positive questions throughout this book. In addition, in Chapter 5 we will discuss in more detail how the IPCC's authoritative status was gained and how it can be defended and enhanced.

Positive questions are not all there is to a policy debate. Policy debates also involve various normative questions, such as what kind of world we want to live in; how we evaluate different kinds of risks and costs, including our attitudes to uncertainty; how we trade off present against future harms; how optimistic we are about the potential for future technological change to ease the problem; and how we value the distributional tradeoffs associated with alternative policy choices. In contrast to positive questions about the climate, these questions are best dealt with through public deliberation and democratic decision processes. Indeed, the widespread pretense that current disagreements over climate-change policy basically arise from disagreements about the state of scientific knowledge has allowed policy-makers to avoid dealing with their real responsibility, which is to engage these questions of political values in view of the present state of scientific knowledge, in order to decide what to do. The distinction between positive and normative questions cannot always be drawn perfectly cleanly in practice, but we contend that it is possible to disentangle these much more than has been done in the present debate, and that efforts to do so are likely to yield a more informed and less contentious policy debate, perhaps even assist in the identification of widely acceptable policy choices.

The remaining chapters follow our ambition to distinguish positive from normative questions. Chapter 3 identifies the most important positive questions about the climate for the present policy debate and summarizes the present state of scientific knowledge about them. Chapter 4 examines alternative policy responses to the climate-change issue, concentrating on present knowledge about what options are available, how effective they are likely to be, and what their costs and other consequences are likely to be. Chapter 5 returns to our central concern about partisan distortion and misuse of scientific knowledge and uncertainty in the policy debate. We outline a few prominent instances of such argument, and discuss in more detail how effective scientific assessment processes could reduce the latitude and the incentives for such misrepresentation. Finally, we step partway into the role of advocates ourselves, and provide a broad outline of an approach to climate-change policy that, in our view, holds the hope of breaking the present deadlock.

Further reading for Chapter 2

Bimber, B. (1996). *The Politics of Expertise in Congress: the Rise and Fall of the Office of Technology Assessment*. Albany: SUNY Press.

> This study examines the relationship between scientific expertise and political decision-making, based on the history of the Office of Technology Assessment (OTA), an office established in 1972 to provide expert scientific and technical advice to the United States Congress.

Jasanoff, S. (1990). *The Fifth Branch: Science Advisors as Policymakers*. Cambridge: Harvard University Press.

> This study of scientific advisory bodies to US government agencies examines the processes by which the boundaries between the scientific and political domains were negotiated in several regulatory controversies, and the conditions that contributed to more or less stable and effective maintenance of the boundary – and a more or less constructive relationship between scientific advice and regulatory decision-making – in each case.

Kuhn, T. (1962). *The Structure of Scientific Revolutions*. Chicago: University of Chicago Press.

> This study of the social processes by which scientific disciplines make progress was the first to note the contrast between two starkly different modes of change in

scientific understanding: normal, incremental progress that depends on certain deep and unexamined structure of shared assumptions (which Kuhn called "paradigms") about what questions are important and what lines of research are interesting or promising; and occasional revolutionary upheavals that follow the accumulation of some critical quantity of results that do not fit within the paradigm.

Mazur, A. (1973). "Disputes between experts". *Minerva: a review of science, learning, and policy*, **11**: 2, April, pp. 243–262.

This study of two past policy controversies marked by sharp public disagreement between scientific experts – fluoridation of drinking water and exposures to low-level radiation – examines how advocates' debating tactics led to public confusion about what was known and increased polarization of the policy debates.

Ruckelshaus, W. D. (1985). "Risk, science, and democracy", *Issues in Science and Technology*, **1**: 3, Spring, pp. 19–38.

Ruckelshaus, Administrator of the US Environmental Protection Agency under Presidents Nixon and Reagan, argues that effective environmental policy depends on maintaining enough separation between scientific-based processes of assessing environmental risks, and more political processes of deciding what to do about the risks based on the best scientific information that is available.

Weinberg, A. M. (1972). "Science and trans-science". *Minerva: a review of science, learning, and policy*, **10**: 2, April, pp. 209–222.

A forceful argument that scientific research cannot always answer the highest-priority questions for policy decisions with the precision, confidence, and timeliness that policy-makers want, even when these are precisely specified positive questions that should in principle be amenable to scientific inquiry.

3

Climate change: present scientific knowledge and uncertainties

This chapter summarizes what we know about climate change, and where the key uncertainties and gaps in our present knowledge lie. Contrary to the impression you might get from following the debate in the news, we actually know a great deal about the climate – about its present status, observed variation and trends, the extent of human influence on it, and potential future changes. We parse the questions of the reality and importance of climate change into four separate, specific questions.

- *Is the climate changing?*
- *Are human activities responsible for the observed changes?*
- *What are the likely climate changes over the next century or so?*
- *What will the impacts of future climate changes be?*

For each of these, we will review the available evidence and summarize the present scientific consensus, the degree of uncertainty, and the key remaining disagreements.

3.1 Is the climate changing?

To answer this question, we must first sharpen it in three ways. First, we must define what we mean by "climate." Climate is not just temperature, but also includes such factors as humidity, precipitation, cloudiness, and winds, etc. Although changes in any of these quantities can matter, we focus on temperature because it is the climatic characteristic for which the best data are available and the one that should be most directly influenced by greenhouse-gas emissions. Second, we must specify the time period we will consider. For human-induced climate change, the relevant time period is the few centuries since the industrial revolution, because this is when human activities have been significantly increasing the abundances of greenhouse gases in the atmosphere. Finally, we must specify where we will look for climate change. We will look for changes in the average surface temperature of the Earth, averaged over the entire year. This is the place where any temperature trend we may see is most reliable, because smaller-scale regional variations tend to average out.

To determine if the Earth's surface is warming, we need measurements of temperature or some related quantity over a long enough period to establish a trend. There are many different sources of relevant data to draw on. None of these is perfect. Each has distinct strengths and weaknesses, and some are more reliable than others overall. We will review several of the most important of these data sources, and will see that they paint a consistent picture of rising temperatures. Considered together, these sources provide decisive evidence that the Earth's surface has been warming over the past century, with particularly rapid warming over the last few decades of the twentieth century. In addition, there is some evidence that the warming extends back several centuries.

3.1.1   The surface thermometer record

The simplest way to measure the temperature of the Earth is to place thermometers – such as simple liquid-in-glass thermometers like the one you may have on your back porch – in many locations around the world, and record the temperature at each location every day. By combining measurements taken at locations all over the globe, you can construct an estimate of the average surface temperature of the Earth. People have been making these measurements at thousands of points over the globe, both on land and from ships at sea, for about 150 years. This combined record shows that during the twentieth century, the global-average surface temperature of the Earth increased by 0.4–0.8 °C (Figure 3.1). Most of this increase occurred in two distinct periods, from 1910 to 1945 and from 1976 to the present, with a small cooling between these periods and with many short-term bumps and wiggles throughout the century. (We will discuss the origin of the cooling period, and of the bumps and wiggles, in Section 3.2.) The 1990s were the warmest decade since measurements began in the mid-nineteenth century, and the warmest individual years (in order) were 1998, 2002, 2003, 2004, and 1997.

Note that Figure 3.1 plots "temperature anomaly" rather than the actual temperature. The temperature anomaly is the difference between the actual temperature each year and some reference temperature. In this figure, the reference temperature is the Earth's average surface temperature between 1961 and

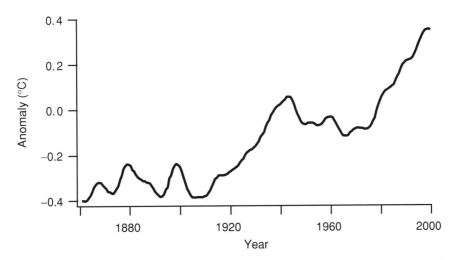

Figure 3.1. Combined annual land-surface air and sea-surface temperature anomalies (°C) from 1861 to 2000, measured relative to the 1961–1990 average. Data have been smoothed to show decadal variations. Source: Figure 1a of the Summary for Policymakers, IPCC (2001a).

1990, about 14 °C. The figure tells us that temperatures between 1860 and 1920 were 0.2–0.4 °C below the 1960–1990 average, while temperatures over the past 10–20 years have been 0.5 °C above this average.

Why show "anomalies," rather than the actual temperature measurements? The main reason is that many sources of temperature data, such as the glacier data described in the next section, can only measure *changes* in temperature over time – equivalent to temperature anomalies – not absolute temperature. Because of this, global temperature data are almost always expressed as anomalies, even if they could be expressed as absolute temperatures, so that the records from all data sources can be compared.

This temperature record provides the strongest evidence that the Earth is warming, as well as the most accurate estimate of how much it has warmed. Why is this data set so good? The primary reason is that these data are the most direct measurements of the Earth's temperature. Other methods of determining the trend in surface temperature are *indirect*. They do not measure surface temperature itself, but infer it from some other quantity such as the size of glaciers or the extent of sea ice. For these indirect data sets, converting changes in the observed quantity to a surface temperature trend introduces additional uncertainty. Another advantage of the surface thermometer record is that the technology behind thermometers is hundreds of years old and we understand exactly how they measure temperature. Such technical maturity adds great confidence that the temperature trend observed in the data really represents a warming, not some undiscovered arti-fact of the instrument being used. Because of these advantages, this data set is the most studied and most trusted in climate-change science. No other data set we will discuss in this chapter is either as well understood or as relied upon in the climate-change debate.

Despite its strengths, this data set still has imperfections. The 150-year his-tory of continuous observations is in some respects a strength, but changes in how observations were made over that long period can also introduce errors. To illustrate the kind of errors that can occur, consider a hypothetical temperature station that has operated from 1861 to the present. In 1861, it was operated by a farmer, who read a liquid-in-glass thermometer and recorded the temperature every day at noon. While the technology of the thermometer was mature even then, there were occasional errors in the record, because of instrument problems (for example, a bubble in the thermometer), or because the farmer mis-read the thermometer or wrote the temperature down incorrectly. Simple errors like these turn out to be relatively unimportant for discerning a long-term trend, because they are no more likely to go in one direction than the other. As a result, they average out in the long term.

When the farmer died in 1890 his son continued the daily temperature read-ings, but he made them at 3:00 in the afternoon instead of at noon. Since it is usually warmer in mid-afternoon, the temperatures recorded at this station sud-denly jumped upward. In 1902, the barn next to the thermometer burned down and the thermometer was moved to a south-facing hillside that received more sun-light, and the recorded temperatures increased once again. Over the next 50 years, the nearby city grew until it eventually surrounded the farm. Cities are warmer than the surrounding countryside, because roads and buildings are darker than vegetation and so absorb more sunlight – a phenomenon known as the "urban heat island effect" – so this urban sprawl caused an additional warming trend in the record. These errors, unlike simple reading and recording errors, can introduce spurious trends in the temperature record.

These types of error are well known, and various techniques are used to iden-tify and correct them. For example, changes in observing practices (for example, changing the measurement time from noon to 3 p.m., or moving the thermometer) can be identified by looking for sudden jumps in a station's temperature, then checking the station's log books to see what changed on that day. Once the cause is identified, the station's prior records can be adjusted to account for the change in observing practices. The size of the urban heat island effect can be estimated by comparing a station in a growing urban area with a nearby rural station. While the urban heat island effect can be important in estimating local or regional trends, it is not a major factor in the global trend shown in Figure 3.1: the trend calculated using only rural stations is very similar to that calculated using all observing stations.

The final problem with the surface thermometer temperature record concerns how thoroughly and uniformly the observing stations cover the Earth's surface. The coverage is extensive, but far from complete. Most stations are located where people live or travel, so most measurements are made on land, in densely populated regions. Coverage is thin over the polar regions, uninhabited deserts, and ocean regions far from major shipping lanes. In addition, coverage has changed over time, especially on the ocean. If the newly added regions are on average warmer or cooler than the regions previously observed, this could also create a spurious trend. As with changes in observing practices, scientists are aware of these problems and have developed techniques to determine a robust average temperature from sparse data and to estimate how much bias might still remain in the record. For example, global satellite measurements of sea-surface temperature now make it possible to determine accurately what kind of errors in calculating global-average temperature were caused by the earlier sparse coverage of measurements over the oceans, as well as changes in the oceanic coverage.

The surface thermometer record is the most important historical data set used in studies of climate change. But as Chapter 2 stressed, important scientific claims (for example, that the Earth is warming) must be verified by several independent observations before they are widely accepted. In the rest of this section, we discuss other data sets that provide independent estimates of temperature trends.

3.1.2   The glacier record

In cold regions, such as near the poles or at high elevations in mountains, snow that falls during the winter does not all melt during the following summer. Under the right conditions, the snow can accumulate to great thickness over many years, compressing under its own weight to form a thickened sheet of ice known as a glacier. Glaciers presently cover about 10 percent of the Earth's total land area, mostly in Antarctica and Greenland. If the climate warms, glaciers will melt and consequently get smaller or "retreat."

Glacier lengths have been measured for hundreds of years, so we can readily determine if they have been melting. Analysis of historical records shows a clear pattern of receding glaciers. Of the 36 individual glaciers monitored over the period 1860–1900, only one advanced and 35 retreated. Of the 144 monitored over the period 1900–1980, two advanced and 142 retreated. Figure 3.2 shows the average change in length of the world's glaciers since 1700. Mean glacier length began declining around 1800, with the decrease accelerating gradually over the

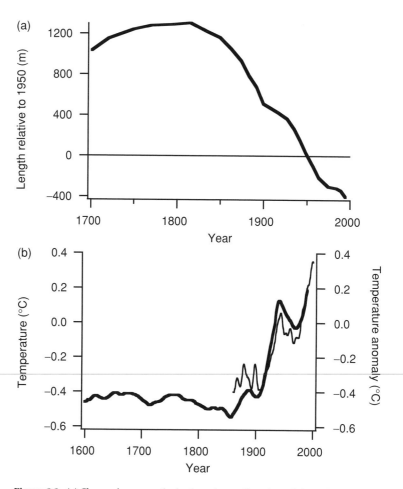

Figure 3.2. (a) Change in mean glacier length as a function of time. (b) Reconstructed global mean temperature inferred from this change in glacier length (thick line, left-hand axis). Also shown is the temperature anomaly based on the surface thermometer record from Figure 3.1 (thin line, right-hand axis). After Figs. 2b and 3b of Oerlemans (2005).

first half of the nineteenth century. Individual regions of the Earth show a decrease that is similar to the global average plotted in Figure 3.2a, reflecting the fact that glacier retreat on the century time scale is rather uniform over the globe.

The most obvious explanation for this widespread retreat of glaciers is a warming climate. Using a simple model of glacier melting, one can infer a warming from the glacier data that is close to the warming in the surface thermometer record (Figure 3.2b), providing an independent verification of the warming seen in those data.

Global warming is not the only possible explanation for the observed reduction in glacier length. A decrease in cloudiness, allowing more sunlight to reach the glaciers' surface, could cause increased melting, leading to a similar decrease. Alternatively, since the expansion or retreat of a glacier is determined by the balance between snow accumulation and melting, a decrease in snowfall could also cause glaciers to retreat, even with no increase in temperature. Models of glacier formation and evolution, called "mass balance models," can calculate the changes in glacier length that we would expect from each of these climatic changes. It would take a 30 percent decrease in cloudiness or a 25 percent decrease in annual snowfall to cause the same retreat for a typical mid-latitude glacier as a 1 °C warming. Such a large change in cloudiness or snowfall could occur locally or even regionally, but worldwide trends this large over a century are unlikely. For this reason, most glaciologists consider a warming trend to be the dominant cause of the observed worldwide glacier retreat.

One limitation of glacier data is coverage. Glaciers are found only in cold places, so a temperature trend calculated from glacier retreat tells only part of the story of worldwide temperature trends, even though it is consistent with the trend in the surface temperature record. We will see below, however, that other data sets with different regional coverage provide similar evidence of warming trends, giving additional support to the warming seen in the surface thermometer record and in glacier retreat.

3.1.3   Sea level

As the Earth's climate warms, the sea level rises, for three principal reasons. First, like most substances, water expands when it warms, so climate warming increases the volume of the water in the oceans. Second, when warming melts glaciers or other ice on land, the melt water runs into the oceans and further raises their level. The opposite effect occurs during ice ages. At the peak of the last ice age, the immense volume of water stored in continental glaciers lowered sea level 120 m below the present level. Third, changes in the amount of water stored on land in forms other than ice, for example in lakes and aquifers, can also change sea level, particularly when ground water is pumped out of aquifers for irrigation or other uses and flows to the ocean.

Data from tide gauges show that over the twentieth century, global average sea level rose by about 1.5 mm per year, or 15 cm in total over the century. The few tide-gauge records that extend back into the nineteenth century suggest that the sea level rose faster in the twentieth century than in the nineteenth century.

Non-climate processes can also affect sea level, complicating attempts to infer a temperature trend. For example, local sinking of coastal land can make local sea

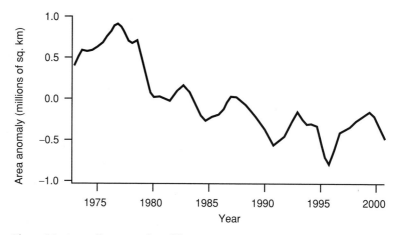

Figure 3.3. Annually averaged satellite-measured Arctic sea ice area anomalies, 1973–2000, relative to the 1973–1996 average. Source: Fig. 2.14, IPCC (2001a).

level appear to rise, even if the absolute sea level is constant. Such sinking can arise from slow natural movements of the Earth's crust (for example, due to plate tectonics or glacial rebound), or from human activities such as groundwater extraction. We have good knowledge of where such sinking is happening, however, and so can adjust for this in interpreting local sea-level measurements. Considering all potential causes, it appears that thermal expansion accounts for about half the observed twentieth-century rise. Glacier melting appears to have made a smaller contribution. Changes in water stored on land might have been important, but their contribution is much more uncertain. Overall, the twentieth-century sea-level rise is consistent with the warming trend seen in the surface thermometer record.

3.1.4   Sea ice

Seawater freezes in the polar regions, forming a layer of ice that is typically a few meters thick on the top of the ocean. Because this occurs only in places where the temperatures are cold, the extent of sea ice provides an indication of where such low temperatures are found. And since all projections of climate change due to greenhouse gases suggest that the strongest warming should occur in polar regions, we expect to see evidence of a warming trend reflected in sea ice – and we do.

Figure 3.3 shows the annual average area of Arctic sea ice (plotted as anomalies relative to a reference area). There is a clear long-term trend: the average ice-covered area was about one million square kilometers smaller in the late 1990s than in the mid-1970s. This rate of decrease, about 2.8 percent per decade, is consistent with observed high-latitude warming over the same period. For

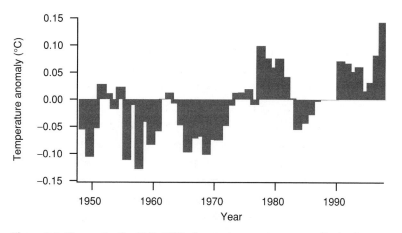

Figure 3.4. Time series for 1948–1998 of ocean temperature anomalies in the upper 300 m for the global ocean. Source: Fig. 2–11, IPCC (2001a).

example, between 1979 and 1998, the length of the Arctic melting season, the period when temperatures are above freezing and the sea ice is melting, increased from 57 to 81 days, while the season cold enough to generate sea ice has decreased.

In addition to shrinking in area, sea ice has also grown thinner. Measurements from submarines show that, over the past several decades, Arctic sea ice has lost about 40 percent of its thickness, decreasing from slightly more than 3 m thick on average to slightly less than 2 m. Together, these sea-ice measurements provide strong evidence of warming in the Arctic region.

These data have clear limits as indications of a global trend. The most obvious limit is that the sea-ice record indicates warming only in the Arctic region, not worldwide. Moreover, the same relationship between warming and sea-ice area does not appear to hold in the Antarctic, where sea-ice area has remained stable since the mid-1970s, or even slightly increased. It is not known whether sea-ice thickness in the Antarctic has changed.

3.1.5   Sub-surface ocean temperatures

In addition to the thermometer measurements of air temperature at the ocean's surface included in the surface thermometer record, temperature measurements have been taken in the upper 300 m (about 1000 ft) of ocean water worldwide since about 1950. Figure 3.4 presents the record of annual temperature anomalies, which shows an average warming trend of about 0.037 °C per decade, a total of 0.18 °C over the 50-year record.

This might seem like a small warming, but because of the large heat capacity of water (something you have experienced if you have ever waited for a cold

Figure 3.5. Photo of the cross-section of a tree, showing the many tree rings. Note that the rings vary in size and density. Photo courtesy of the Laboratory of Tree-Ring Research, University of Arizona.

hot-tub to warm up), it actually represents a large increase in the stored energy of the ocean. When climate models are used to relate this warming of the surface ocean to greenhouse-gas abundances in the atmosphere, they find that the ocean warming is consistent in sign and in approximate magnitude with greenhouse-gas increases over the past few centuries.

3.1.6   Climate proxies

A "proxy" climate record is a record of past climate variation that has been imprinted on some long-lived physical, chemical, or biological system. Because of their longevity, climate proxies can provide evidence of past climate from long before the modern instrumental record. They give a window into how the Earth's climate has varied over the long term, allowing us to ask when, if ever, the Earth has experienced periods as warm as the present or rates of warming as rapid as that of the past few decades. This section discusses several of the most important and widely used sources of climate proxy data.

Tree rings

Tree growth follows an annual cycle, which is imprinted in the rings in their trunks. As trees grow rapidly in the spring, they produce light-colored wood; as their growth slows in the fall, they produce dark wood. Figure 3.5 shows a cross-section of a tree trunk and its rings. Because trees grow more – and produce wider rings – in warm years, the width of each ring gives information about climate conditions around the tree in that year. The rings of a long-lived tree can provide a temperature time series that extends back hundreds of years.

The key to using tree rings as a climate proxy is finding a quantitative relation between tree-ring width and temperature in the tree's location. This is done by examining rings from recent years when thermometer records are also available. Once a relationship between ring width and temperature is estimated for a recent period, this relationship can be used to estimate the temperature for the period before there were direct measurements.

There are two principal difficulties to using tree rings as a climate proxy. First, it is difficult to separate temperature's effects on tree growth from those of other climate characteristics such as rainfall. Second, the method assumes that the relation between tree-ring width and temperature determined from recent data applies over the entire life of the tree. There is no real way to know whether this is so or not. Further, tree-ring reconstructions are available for only a small part of the Earth's surface. They are obviously not available over oceans. Nor are they available from desert or mountainous areas where no trees grow, or from the tropics, because the small seasonal cycle there means that trees grow year-round and so produce no rings.

Ice cores

Both Greenland and Antarctica are almost entirely covered by glaciers hundreds to thousands of meters thick. We discussed above how the advance or retreat of glaciers gives information about temperature changes over the past few centuries, but glaciers can provide more climate information as well. The chemical and physical characteristics of the glacial ice provide a rich store of information about conditions at the time the snow fell. Small air bubbles trapped when glacial ice is formed preserve a picture of the chemical composition of the atmosphere at that moment.

If there are no trapped air bubbles in a particular layer of ice, this means that when the ice formed, summer temperatures were warm enough to melt the top layer of ice. In addition, the chemical composition of the ice, in particular the fraction of heavy isotopes of hydrogen and oxygen (forms of hydrogen or oxygen atoms that contain extra neutrons) can be used to infer the air temperature around the glacier when the snow fell, as can variations in the size and orientation of ice crystals. The amount of dust trapped in the ice conveys information about how wet or dry the regional climate was when the ice formed, because more dust blows around during droughts, and about prevailing wind speed and direction. Finally, sulfur is one of the main effluents of volcanoes. Once emitted to the atmosphere, this sulfur dissolves in rain, forming sulfuric acid which is transferred to the ice

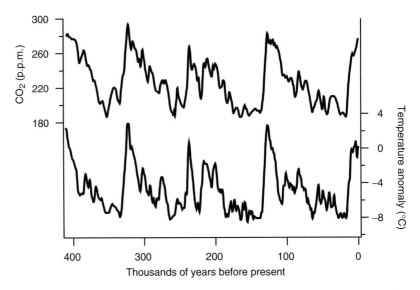

Figure 3.6. Data from the Vostok ice core in Antarctica, from 410 000 years ago to the present. The top curve shows abundance of CO_2 (in parts per million) from air bubbles in the ice core. The bottom curve shows the temperature anomaly in the Antarctic region, relative to the present, from isotopic measurements of the ice. Source: adapted from Petit *et al.* (1999).

when rain or snow falls on the glacier. Measurements of the acidity of glacial ice consequently tell us whether there was a major volcanic eruption around the time the ice was formed.

Researchers retrieve a time series of all this information by drilling down into the ice sheet with a hollow drill bit and removing a long column of ice, a few inches in diameter, known as an ice core. The further down you drill, the older is the ice you retrieve. Reconstructing information about historical climate from an ice core requires two steps. First, the age of each layer of ice must be determined from its depth inside the glacier. Although much effort has been spent on this problem, it still carries important uncertainties, because the rate of ice accumulation varies over time and because ice inside the glacier can compress and flow under the great weight of the ice above. Second, the characteristics actually observed must be translated into the climatic characteristics of interest – for example, translating abundance of heavy water to temperature, and amount of dust to precipitation – introducing additional uncertainties.

Figure 3.6 shows data from an ice core in Antarctica – atmospheric CO_2 concentrations obtained from air bubbles trapped in the ice, and temperature variations calculated from isotopic measurements – that go back an astonishing 410 000 years. Over this period, variations in CO_2 and temperature have been large, and have followed each other very closely. Human activities can have played no conceivable role in variations over most of this time, of course. It is only recently that we have had the industrial might to affect the Earth to any significant extent. Rather, these past variations are driven by small changes in the orbit of the Earth that have changed the climate by modulating the amount of sunlight falling on the Earth (see Section 3.2.1).

It is likely that the temperature variations in Figure 3.6 drove the CO_2 variations, not the reverse. This might have occurred, for example, when warmer temperatures increased the rate of bacterial breakdown of dead plant material, releasing CO_2 to the atmosphere as the atmosphere warmed. This historical relationship does not, however, refute the modern relationship of human additions of CO_2 to the atmosphere driving increases in temperature.

Corals

Corals are small marine animals that live in colonies anchored to reefs in warm ocean waters, mostly in tropical latitudes. The reefs, which are made up of skeletons of previous generations of coral, can be thousands of years old. The chemical composition of the reef can provide information about past climate and ocean conditions. Quantities such as ocean temperature, precipitation, salinity, sea level, storm incidence, and volume of nearby freshwater runoff are all obtainable. As with ice cores, these data give a time series of historical conditions over the life of the reef, subject to two important uncertainties: determining the age of each bit of coral, and converting its chemical make-up to quantities of interest, such as the average ocean temperature.

Ocean sediments

Billions of tons of sediment accumulate at the bottom of the ocean every year. Like ice cores and corals, this sediment contains information about nearby climate conditions when it was deposited. The most important source of information in sediments comes from the skeletons of tiny marine organisms. The ratio of the abundance of species that thrive in warm waters to the abundance of species that thrive in cold water tells us about the surface water temperature. The chemical composition of the skeletons and variations in the size and shape of particular species provide additional clues. In the end, information about water temperature, salinity, dissolved oxygen, nearby continental precipitation, the strength and direction of the prevailing winds, and nutrient availability can all be obtained from ocean sediment.

Boreholes

Temperatures measured today at different depths underground provide a different way to infer how the surface temperature varied in the past. To

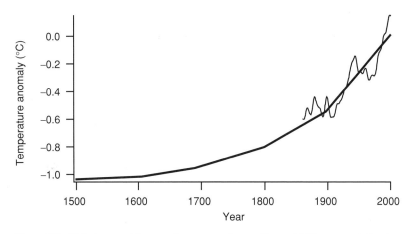

Figure 3.7. Global ground temperature anomaly over the past 500 years, relative to present day, estimated from borehole data (thick line), with the surface thermometer record (thin line), from Figure 3.1. Source: Fig. 2.19, IPCC (2001a).

understand how this works, think about cooking a frozen turkey. You can tell how long the turkey has been in the oven by measuring the temperature at different depths below its skin. If the turkey is hot on the surface but still frozen just below the skin, then it has been cooking for only a short time. If the center of the turkey is 165 °F, then it has been in the oven for several hours – and you should take it out before it is overcooked! In an analogous way, measuring the temperature of the Earth at many depths in deep narrow holes called boreholes allows you to infer the history of the ground surface temperature over the past few hundred years.

Figure 3.7 shows a reconstructed global ground temperature history from several hundred boreholes, most of them in North America and Eurasia but with a few in Africa, South America and Australia. The record shows that average ground temperature has increased by about 0.5 °C during the twentieth century and about 1.0 °C since 1500, and that the twentieth century was the warmest of the past five centuries. The good agreement between the borehole data and the direct surface air temperature data, shown in the thin line, gives increased confidence to both, and to the strong warming trend they both show over the past century.

One issue is that the borehole measurements tell us about the temperature of the ground. The surface thermometer measurements, on the other hand, tell us about the temperature of the air a few meters above the ground. Usually the difference in trends derived from these two sources is small, but depending on the properties of the surface – for example, land-use and land cover, soil moisture, and winter snow cover – significant differences can exist. In central England, for example, where the ground is rarely snow-covered and major land-use changes

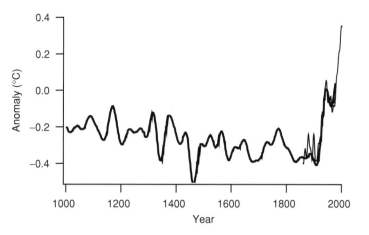

Figure 3.8. Northern Hemisphere temperature anomaly, relative to the 1961–1990 average, from climate proxy data for period 1000–2000 (thick black line). These data have been smoothed to show longer-term variation. The thin line at the right shows the direct surface air temperature record, from Figure 3.1. Source: Fig. 2.20, IPCC (2001a).

have not occurred for several centuries, surface temperature trends inferred from borehole records are very similar to those in thermometer records. But in north-western North America, borehole estimates of surface warming in the twentieth century are 1–2 °C larger than the warming in the thermometer record, most likely because of changes in average snow-cover and land-use during the century. Such discrepancies are considered and controlled to the extent possible in order to construct a consistent historical temperature record.

A combined proxy climate record

There are several other sources of climate proxy data in addition to those we have discussed. Each proxy provides a different view of climate history: for example, ice cores provide information about polar regions, tree rings about mid-latitudes, and corals about the tropics. While each individual data source has unique coverage in time and space, and its own uncertainties, they present a robust and complete picture of the climate when they are combined. Figure 3.8 shows a combined reconstruction, from multiple climate proxies, of Northern Hemisphere temperature anomalies from the year 1000 to 2000. The combined record shows a gradual cooling trend through most of the millennium, with an abrupt warming beginning around 1900. These records suggest that the 1990s were the warmest decade, and 1998 was the warmest year, not just of the past 150 years, but of the past 1000 years. Several research groups have constructed such combined records using different sets of proxies and different methodologies. All have found similar historical climate trends, giving confidence in the results.

3.1.7   Satellite temperature measurements

Meteorological satellites have provided a new, independent source of data about global temperatures since 1979. One type of satellite instrument in particular, the Microwave Sounding Unit (MSU), measures microwave radiation emitted by the Earth's atmosphere, from which it is possible to calculate the temperature of the atmosphere at various altitudes. As a measure of global-average temperature, this satellite data set has the advantage that it covers the entire Earth, including the oceans and uninhabited land areas, and so avoids potential problems from partial or biased coverage. It is also a more direct measure of temperature than some of the sources we have discussed, such as climate proxy data, although less direct than the surface thermometer record.

Despite its important advantages, the MSU data also have some critical weaknesses. First, the observations cover only 25 years, a rather short period to draw strong conclusions about trends. Even more seriously, this record of just 25 years is constructed out of data from 12 separate satellites. To understand why this is a serious problem, suppose you are keeping track of your weight to tell if you are gaining or losing. Further, suppose your scale breaks, and a month passes before you buy a new one. If the new scale says you are 5 pounds heavier than your last reading on the old one, does this mean you have gained 5 pounds? Or does the new scale just weigh *everything* 5 pounds heavier than the old one? You could avoid this problem by buying a new scale *before* the old one breaks, and measuring yourself on both scales for a while to estimate the difference between them – if you had the foresight, patience, and money to do this.

The MSU record suffers from the same problem, because each satellite only lasts a few years. The agency that operates them (the US National Oceanographic and Atmospheric Administration, or NOAA, which includes the National Weather Service) tries to launch each new satellite while the previous one is still operating, to provide a long enough period of overlapping measurements for comparison. But since you cannot predict precisely when an instrument is going to fail, NOAA has not been entirely successful in obtaining long enough overlapping records. In particular, the NOAA-9 satellite had only a short overlap with the NOAA-7, -8, and -10 satellites, and the temperature trend estimated from the MSU data is quite sensitive to how you connect data from satellites that flew before NOAA-9 to those that flew after.

Figure 3.9 shows global-average temperature trends calculated from the MSU data by two scientific groups, together with the trend in the surface thermometer

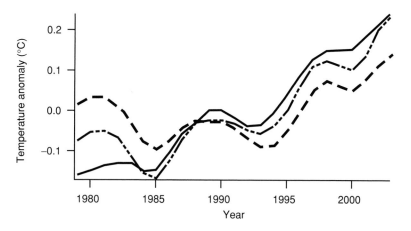

Figure 3.9. Temperature anomalies calculated from MSU satellite data and the surface thermometer record over the period 1979–2003. The dashed line shows the MSU anomalies calculated by the University of Alabama in Huntsville (Christy *et al.*, 2003); the dot-dot-dashed line shows the MSU anomalies calculated by Remote Sensing Services in Santa Clara, CA (Mears *et al.*, 2003). The solid line shows the anomaly in the surface thermometer record over the same period. Source: the NOAA National Climatic Data Center (http://www.ncdc.noaa.gov/oa/climate/research/2003/ann/global.html).

record. There is a large difference between the MSU trends calculated by the two groups. The University of Alabama in Huntsville (UAH) group estimates a warming trend of 0.06 ± 0.05 °C per decade (see Christy *et al.*, 2003)[1] – substantially smaller than the 0.1–0.2 °C per decade trend in the surface thermometer record over the same period – while the Remote Sensing Services (RSS) group estimates a warming trend of about 0.15 °C per decade (Mears *et al.*, 2003) which is close to the surface record.[2] A third group has recently published its estimate of an even stronger warming trend in the MSU data, 0.22–0.26 °C per decade (Vinnikov and Grody, 2004).

The differences in the satellite warming trends calculated by each group come from different assumptions about uncertain technical details of the trend calculation, for example how to handle the satellite-to-satellite calibration problem described above. It has also been recently suggested (see Fu *et al.*, 2004) that none

---

[1] Note that an earlier analysis by the UAH group indicated that the atmosphere was actually cooling. Newer calculations by the UAH group that incorporate a better understanding of the data as well as a longer time series now show the small warming trend stated here.

[2] Mears *et al.* (2003) state that the RSS-calculated trend is 0.09 °C per decade larger than the UAH trend. The value stated here, 0.15 °C per decade, is obtained by adding 0.09 °C per decade to the UAH trend of 0.06 °C per decade.

of these calculations adequately accounts for how recent cooling of the strato-sphere affects the MSU measurements.[3] Correctly accounting for this effect might increase the calculated warming trends in the troposphere by 0.05 °C per decade. There are also several other known but unresolved problems with the MSU data. For example, strong correlations between each satellite's measurements of the atmospheric temperature and of its onboard calibration target strongly suggest that there are still unresolved errors in the calibration of the satellite instruments.

While the trends calculated from MSU data span a wide range, they are unan-imous in finding a significant warming. The disagreements among calculated MSU trends, and between the MSU trends and the surface record, only concern how much warming is occurring, not whether or not warming is occurring. The large discrepancy between the surface trend and the satellite trend calculated by the UAH group in particular has been used prominently in the policy debate to cast doubt on the claim that the Earth is warming – a controversy that we will discuss further in Chapter 5. The discrepancies between the various trends, together with the known weaknesses of the MSU data, indicate that these data cannot at present provide a precise estimate of warming of the lower atmosphere but only a rela-tively wide range, from near zero warming to about 0.3 °C per decade. Interpreting these data is an active area of current scientific research, and it is likely that future research on this data set will allow this wide range to be substantially reduced.

3.1.8   Summary: is the Earth warming?

Table 3.1 summarizes what we know about trends in the Earth's temper-ature. All this evidence has been peer reviewed and multiply verified by indepen-dent scientific groups. There are more sources of relevant evidence that we have not discussed, but this list includes the most important data sets discussed in the Working Group I report of the IPCC *Third Assessment Report*. An examination of Table 3.1 is striking: every source of relevant data shows the Earth has warmed over the past century or so. There is also some evidence that the warming has been underway for several centuries, but the evidence for this longer trend is mixed.

No data set is perfectly reliable, of course. It is possible that any one of these data sets could be significantly in error, although the critical scrutiny and mul-tiple verifications that each of them has received minimizes this risk. But there is essentially no chance that enough of these data sources could be wrong by far enough, and *all in the same direction*, that the overall conclusion of substantial

---

[3] The stratosphere is warmed by the absorption of sunlight by ozone. Depletion of ozone over the last few decades has resulted in a strong cooling of the lower stratosphere.

Table 3.1. *A summary of measurements of changes in the Earth's temperature*

Type of data	Direction of twentieth-century change	Size of change, comments
Direct surface air temperature	Warming	Average surface air temperature increased about 0.6 °C (1 °F) over the twentieth century, with about half this warming occurring between 1980 and 2000.
Glaciers	Warming	Glaciers have been receding on average for a few centuries, with evidence of faster retreat in the twentieth century. The warming implied by this recession is about two-thirds of a degree Celsius per century, consistent with the surface record.
Sea-level change	Warming	Sea level rose about 15 cm total over the twentieth century. About half this rise probably came from the expansion of ocean water as it has warmed.
Sea ice	Warming	The area of Arctic sea ice in spring and summer has decreased by 10–15 percent over the past 50 years. Average thickness of Arctic sea ice has decreased by 40 percent over the same period.
Ocean temperature	Warming	The top 300 m of the ocean has warmed 0.18 °C over the past 50 years.
Climate proxies	Warming	Combined data from many climate proxies shows gradual Northern Hemisphere cooling from the year 1000 to the nineteenth century, then strong warming during the twentieth century.
Satellite temperature measurements	Warming	Satellite instruments show warming of 0.06–0.26 °C per decade

global warming in the twentieth century could be wrong. Under the weight of this abundant, consistent, thoroughly checked evidence, the relevant scientific community has overwhelmingly accepted the claim that the Earth's surface has warmed over the past century.

3.1.9   What is NOT evidence that the Earth is warming

The Earth is warming. But not everything we see that is consistent with warming gives additional support to this conclusion. This may seem paradoxical, but it is not. The distinction between "being consistent with" warming and "providing additional support for" warming is best illustrated by one-time or regional events such as the disintegration of the Larsen B ice shelf in 2002. An ice shelf is an ice mass floating on the ocean, formed when a glacier flows into the ocean and extends away from the shore like a floating dock. Over a one-month period in early 2002, the northern section of the Larsen B ice shelf – a mass of ice about 250 meters thick covering more than 2500 square kilometers – spectacularly shattered, creating thousands of enormous icebergs.[4] Scientists agree that the breakage was caused by warming *in the region around the ice sheet*, but what caused this regional warming? We simply do not know. It is possible, perhaps even probable, that global warming played a role. But regions of the Earth, even regions as large as the United States, can experience warm or cool periods even in the absence of any global trend. Look around the world at any time, and you will almost always see a heat wave going on somewhere, and a cold spell going on somewhere else. You cannot infer a global trend from such local extremes, because local or regional behavior can be different from the global average. The same is true for short-term events. As we discussed above, many glaciers have had periods of growth lasting a few decades over the past few centuries, within a longer-term, worldwide trend of substantial glacier retreat. If you looked at a single glacier during one of these growth periods, you might conclude that the Earth was cooling, but this would be erroneous.

The point of this distinction is not to read too much into regional events or short-term trends, however dramatic these may be. You can say that discrete events like the Larsen B collapse are "consistent with" global warming, but such a single event, by itself, adds little to our confidence that the Earth is warming. The conclusion that the Earth is warming sits on the much stronger foundation of many independent pieces of evidence, over the entire world, over many decades or centuries.

3.2   Are human activities responsible for global warming?

The second question we consider concerns the causes of the observed climate change: *are human activities responsible, or might recent warming be caused by some natural process?* This question is harder to answer than the question of whether the Earth is warming, since establishing a cause-and-effect relationship requires an inference that merely identifying a trend does not. Showing human causation

---

[4] See Revkin, A. C. "Large ice shelf in Antarctica disintegrates at great speed", *New York Times* (Late edition (East Coast)), March 20, 2002, p. A13.

requires both demonstrating that human emissions can account for the observed warming trends, and showing that other potential explanations cannot.

While human emissions are an obvious potential cause of twentieth-century warming, it is entirely reasonable to question whether they really are responsible. Over the history of the Earth, the climate has undergone large fluctuations without any human influence. About 100 million years ago, during the age of dinosaurs, the Earth was so much warmer than today that there was no ice or snow at either the North or South Pole, and plants that today live only in the tropics flourished at high latitudes. On the other hand, 20 000 years ago (at the peak of the most recent ice age) the Earth was so much colder than today that ice sheets thousands of feet thick covered much of North America. These climate fluctuations took place long before human activities could have played any significant role in the changes. But over the past few centuries, human activities have expanded to the point that they can significantly influence many global-scale processes, so they must be considered a potential cause of the observed global warming.

In this section we examine available knowledge about the six potential causes that have been proposed for twentieth-century warming, including human emissions and five natural processes. We find that, for the last half of the twentieth century at least, human emissions of greenhouse gases very likely account for the great majority of the warming.

3.2.1   Orbital variations

It has been known since the Renaissance that the Earth's orbit is not a perfect, unchanging circle, but an ellipse whose shape and orientation change slowly over time. There are three important types of orbital variation. First, the average distance between the Earth and the Sun slowly increases and decreases, completing a cycle every 100 000 years. Second, the time of year when the Earth is closest to the Sun varies. At present, the Earth is closest to the Sun during Northern Hemisphere winter, but in 10 000 years the closest approach will be in Northern Hemisphere summer. Third, the tilt of the Earth's polar axis relative to the Sun, which is now about 23°, slowly oscillates between about 22° and 25° over a period of 40 000 years. The variation in average Earth–Sun distance changes the total amount of solar energy reaching the Earth. Since the climate is driven by solar energy, this variation clearly can change the climate. The other two forms of variation do not change the total sunlight reaching Earth, but change its distribution during the year and over the Earth's surface. For example, variation in the Earth's tilt alters how much sunlight falls on the tropics relative to the polar regions. Such changes in the distribution of sunlight can also affect the climate.

It is now widely agreed that these slow orbital variations cause the cycling between ice ages and warm interglacial periods that the Earth has experienced over the past few hundred thousand years (see Figure 3.6). This conclusion is based on the near-perfect agreement between the timing of the orbital variations and of the observed climate change.

So if orbital changes caused the climate changes of the past few hundred thousand years, could they also be causing the warming of the past century? They almost certainly cannot, because these orbital wobbles are so slow that it takes thousands of years for them to make any significant change in the pattern of incoming sunlight. The warming of the past century has been much too fast to have been caused by these slow orbital variations. The warming must be due to other causes.

3.2.2   Tectonic activity

Tectonic processes, the geological processes that control the distribution of continents and mountain ranges on the Earth's surface, are a second potential natural cause of climate change. Changes in the arrangement of continents and mountains can change the climate in several ways. For example, the location of the continents determines how much land area is covered by snow. Snow strongly reflects incoming sunlight, so when more of the Earth's surface is snow-covered, more incoming sunlight is reflected by the surface and less is absorbed, cooling the climate. Consequently, if continents were to move toward the equator, less land would be snow-covered and the climate would warm. Changing the distribution of continents and mountains can also alter precipitation patterns. Because chemical reactions between rainwater and exposed rock remove CO_2 from the atmosphere, changes in how much rain worldwide falls on exposed rock can change the climate by changing how much CO_2 is in the atmosphere. A powerful illustration of this effect occurred 40 million years ago, when the Indian subcontinent collided with the Asian continent to form the Himalayas and the adjacent Tibetan Plateau. (Collisions between continents happen slowly: this one is still going on today.) The prevailing winds of the time brought heavy rainfall onto the newly exposed rock of these geological features, and the resultant chemical weathering drew down atmospheric CO_2 and caused the Earth's climate to cool over the next 30 million years.

Could tectonic processes have caused the observed warming of the past century? As with orbital variations, they cannot have, because they are much too slow. The cause of the recent warming must be able to affect temperature over a period of a century or two, or even faster. Because tectonic processes take millions of years to move a continent or form a mountain range, they simply cannot do this.

3.2.3   Volcanoes

Volcanic eruptions can change the climate by blowing dust and ash into the atmosphere. The dust and ash block incoming sunlight, cooling the Earth for several years after a major eruption. In 1816, for example, after three major eruptions in three years, the northeastern USA experienced the famous "year without a summer" (see, for example, Stommel and Stommel, 1983). Snow fell in Vermont in June and summer frosts killed many crops, leading to widespread food shortages. When that summer was followed by a winter so cold that the mercury in thermometers froze (this happens at $-40\,°C$), many residents fled the Northeast and moved south.

Could volcanic eruptions somehow account for the observed warming? This is extremely unlikely because of the short time over which volcanoes affect the climate. Within a few years of an eruption, the dust settles out of the atmosphere and the climate returns to normal. The observed temperature trend could be generated by a reduction in atmospheric dust sustained over most of the past century. But this would require a series of massive eruptions every few years, each one precisely calibrated in its timing and magnitude. We have good records of volcanic eruptions over the last century or two, and while they appear to account for some of the bumps and wiggles in the global temperature record shown in Figure 3.1, there is no sign of a sustained pattern of eruptions resembling that required to explain the observed warming. As a result, we can safely rule out volcanoes as a source of the observed trend.

3.2.4   Solar variability

Because sunlight is the power source that drives the climate, any change in the amount of sunlight reaching the surface can change the climate. For example, orbital variations and volcanoes affect the climate by modulating the amount of sunlight reaching the surface. There is also variation in the power output of the Sun itself: it does not shine with constant brightness, but flickers like an old light bulb (it is 5 billion years old, after all). We do not notice this flickering, because it occurs slowly, over periods of months, years, and possibly longer, and because it only changes the Sun's total energy output by a few tenths of a percent. But this variability is large enough to affect the climate. If the Sun's brightness had increased by enough and with the right timing, then this alone could have caused the observed warming of the past century.

Accurate measurements of the Sun's output have been made from satellite instruments since the late 1970s. Over this period, there has been essentially no trend in the Sun's output, only the periodic variation of less than 0.1 percent that occurs over the 11-year solar cycle. Because of the enormous thermal inertia of the oceans, the climate is quite insensitive to such short-term variations. As a result, when put into a GCM, these solar cycle variations produce only a very small effect, and are unable to reproduce the rapid warming observed of the last few decades of the twentieth century. Consequently, the suggestion that increases in solar brightness caused the observed recent warming can be decisively rejected.

But whether changes in the Sun's output could have contributed to earlier warming is a more difficult question. To estimate solar output before it was measured directly from satellites, we must infer it indirectly from measurements of related quantities, much as we use proxies like tree rings or ice cores to infer past climate conditions. One of the main proxies used for solar output is the number of sunspots, which people have been observing and recording for thousands of years. Using this historical sunspot record, together with a relationship between the number of sunspots and solar output constructed from the past few decades when they have both been measured, it has been estimated that solar output has increased by 0.2–0.4 percent over the last few centuries. When this long-term increase is put into a GCM, we find that solar variability could have warmed the planet over these few centuries by as much as 0.5 °C, with much of this warming occurring in the late nineteenth and early twentieth centuries. Based on this evidence, scientists have concluded that solar variability was probably an important or even dominant driver of climate change up until the mid-twentieth century, but has made at most a very small contribution to the rapid warming of the last few decades.

3.2.5   Internal variability

All the potential sources of warming discussed so far involve "forced variability," by which we mean that the Earth's climate responds to some external change, such as a change in the Earth's orbit or the Sun's brightness. But the Earth's climate system is so complex that it can experience variability even with no changes in the external conditions driving the climate – rather like the wobbling of a spinning top, but much more complex. Such climate variation that is unrelated to any external forcing is called internal variability.

Several prominent patterns of internal climate variability are becoming increasingly well documented and understood. The best known is the Southern Oscillation, a sloshing of warm surface water back and forth across the South Pacific Ocean with a somewhat irregular period of several years. The two phases of the Southern Oscillation, called El Niño and La Niña, each last a year or two. In the

El Niño phase, warm surface water builds up at the eastern edge of the tropical Pacific, so the ocean off the west coast of South America warms dramatically. Linked changes in temperature and rainfall extend worldwide, and the Earth's average temperature increases. The La Niña phase reverses these changes, including a cooling of the Earth's average temperature. Several other characteristic patterns of natural climate variability have now been identified, with periods ranging from a few years to a few decades.

Could such internal variability be responsible for the warming observed over the past century? In other words, could the climate be warming all by itself? To begin to answer this question, we look at climate proxy data from before 1800. Since human activities likely had a minimal impact on the climate before then, that portion of the record gives a good picture of patterns of natural variability in the climate. Between 1000 and 1800 (see Figure 3.8), this record shows nothing similar to the rate and magnitude of warming since the late nineteenth century, so if recent warming is due to natural variability it is of a type that has not been evident for at least 1000 years.

Going back before 1000 years ago, the proxy climate data are lower in quantity and quality, so we know less about how temperature varied from year to year. Over the past few tens of thousands of years, there is some evidence of rapid climate changes, with average temperature changes of up to a few degrees Celsius occurring over a few decades to a century or so, during transitions into or out of ice-age conditions. These rapid natural climate changes, however, occurred together with rapid reorganization of circulation patterns in the atmosphere and oceans. More work on this is needed, but there is no evidence at present to suggest that such large-scale changes in atmospheric or ocean circulation are occurring in parallel with the observed twentieth-century warming.

We can also gain insight into natural climate variability by using computer simulations of the climate, usually called General Circulation Models or GCMs. (See the Aside below for more information about GCMs). When climate models are run without any human greenhouse-gas emissions, they show variations in global-average temperature from year to year and decade to decade that are very similar to those seen in the climate proxy data before about 1850 (Figure 3.8), but they produce nothing resembling the rapid temperature increases of the past century. We will see below that climate models can generate such rapid warming only when they include human greenhouse-gas emissions. Considered together, these pieces of evidence suggest that while we cannot definitively exclude natural climate variability as a contributor to recent warming, it is highly unlikely that natural variability can account for any significant fraction of the recent rapid warming.

Aside: what is a climate model?

Believe it or not, you can get an idea of the job that climate models do (often called General Circulation Models, or GCMs) by thinking about fashion models. It is hard to imagine what clothes will look like on a person if you only see them on a hanger. So fashion designers hire models to wear their clothes at fashion shows, so people can get a better idea of what the clothes will look like when worn.

The fundamental physical laws that govern the behavior of the climate system – for example, conservation of energy, conservation of momentum, conservation of mass – can be written down in a few equations. But much like clothes on a hanger, it is impossible to look at these equations and get a sense of how the atmosphere will behave. So climate scientists use these equations to construct a simulated Earth – a climate model, or GCM – in the computer. You can test the simulated Earth by comparing its behavior to that of the real Earth. And you can study the simulated Earth in ways that you cannot study the real Earth. You can examine how it responds to various "what-if" scenarios – what if the output of the Sun changed, or what if there were no human emissions of CO_2 – and compare the results to the real climate behavior we have observed, to test various causes proposed for past climate trends. A good example of this will be shown in the next section. You can also use a model to predict how the climate will respond to any specified assumption about trends in future emissions of CO_2 and other greenhouse gases.

Unfortunately, the atmosphere is too complex to be represented exactly by any present-day computer, so all climate models make approximations and assumptions to be tractable. The most important simplification concerns the smallest size at which atmospheric processes are represented. Most climate models divide the atmosphere into boxes about 100 kilometers square and one or two kilometers thick vertically, and assume that within each box, all conditions (temperature, humidity, winds, etc.) either are uniform, or can be described by a simple mathematical relationship. Understanding the errors introduced by this approximation, and reducing them by improving the mathematical representations of finer-scale physical processes (called "parameterizations") is an area of great effort and the source of some of the greatest controversies in climate science.

To check the validity of climate models, scientists examine how well they reproduce a climate period for which we have good data. For example, they might start the model in the year 1500, run it to the present, and examine how well it reproduces the actual climate record. Present models do quite a good job of simulating the global average historical climate record for the world as a whole and for large continental regions. Simulations agree less well with the instrumental record for smaller-scale regions, however, suggesting that we should have less confidence in future climate predictions as we look at smaller scales.

3.2.6   Human activity

The last potential explanation for the observed warming of the Earth is human activity. There are several reasons to think that this can account for some portion of the observed warming. We know that human activities have been increasing the concentration of CO_2 and other greenhouse gases in the atmosphere for at least the past century or two. Measurements show the concentration of CO_2 has increased about 30 percent over that time (Figure 1.1), while other greenhouse gases have increased by similar or larger amounts. Basic physics provides strong theoretical reasons to believe that such an increase in greenhouse gases should warm the Earth. In addition, there is a rough match between the timing of the observed warming and the buildup of CO_2 over the past 100–200 years, although the match is not perfect. For example, the buildup of greenhouse gases alone cannot explain the slight global cooling that occurred between about 1945 and 1975 (Figure 3.1). Finally, several aspects of the spatial and seasonal pattern of observed warming match what we would expect if the warming were caused by greenhouse gases, including greater warming at high latitudes than at low latitudes and more warming in winter than in summer. As we will show in the next section, there are strong reasons to believe that greenhouse gases emitted by humans are responsible for most of the warming in the second half of the twentieth century.

3.2.7   Summary: are human activities responsible for recent warming?

We have considered six potential causes for the observed warming of the Earth over the twentieth century: human emissions, and five natural processes. Of the five natural processes, two – orbital variations and tectonic processes – can be decisively eliminated as significant contributors to the twentieth-century warming. They are simply too slow to cause significant warming over time periods as short as a century. The other three natural processes – volcanic eruptions, changes in solar output, and internal variability of the climate system – all might have contributed to climate variation or warming over the entire twentieth century, but the evidence summarized above suggests they are unlikely to have contributed more than a small fraction of the rapid warming of the past few decades.

All these potential causes can be evaluated and compared more precisely by using climate models. A model can test various causes by running it several times, each time including a different set of potential climate drivers, and comparing how well each run reproduces the observed climate.

Simulated annual global mean surface temperatures

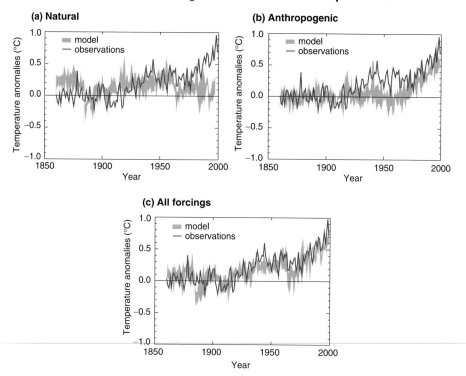

Figure 3.10. Global mean surface temperature anomalies from the surface thermometer record (thin line on all plots), compared with a coupled ocean–atmosphere climate model (thick line). (a) Model includes solar and volcanic effects only. (b) Model includes human greenhouse-gas emissions, aerosols, and ozone depletion, but no solar or volcanic effects. (c) Model includes solar, volcanic, and human effects. Anomalies are measured relative to the 1880–1920 mean. Source: Fig. 12.7, IPCC (2001a).

Figure 3.10 shows such a comparison, between the observed surface temperature record since 1860 and a climate model using three alternative combinations of climate forcing factors. The calculation in panel (a) includes solar variability and volcanoes, but no human effects. This calculation captures some of the bumps and wiggles in the temperature record, suggesting that solar and volcanic effects are indeed affecting the climate. But there are also large differences: in particular, the model does not capture the rapid warming observed since 1970.

The model in panel (b) includes the effects of human activities – greenhouse-gas emissions, and also sulfur emissions from burning coal and stratospheric ozone depletion, both of which tend to cool the surface – but no solar or volcanic effects. This model captures the rapid warming observed since 1970, as well as the slight global cooling that occurred between 1950 and 1970. The sulfur emissions are particularly important to include because sulfur reacts with water vapor in the atmosphere to form small droplets called aerosols, which reflect incoming sunlight and cool the surface. The mid-century period of cooling can be attributed to increased reflection of sunlight from increasing sulfur emissions. Although CO_2 emissions were increasing at this time, the cooling effect from sulfur-based aerosols increasing the reflection of sunlight dominated the warming effect from increased CO_2. Since the 1970s, rapidly increasing CO_2 emissions coupled with slowing growth of sulfur emissions mean that warming from CO_2 now dominates the cooling from aerosols, leading to the rapid warming seen over that period. This model captures this mid-century cooling and rapid, late-century warming, in good agreement with the data. This model run does not, however, capture either the bumps and wiggles in the data or the warming in the early twentieth century, which the previous model run captured better.

The model in panel (c) includes both human emissions and solar and volcanic effects. This model captures all the large-scale features of the historical record: the early twentieth-century warming (mostly due to solar effects), the mid-century cooling (mostly due to sulfur emissions), and the rapid warming of the past few decades (due to greenhouse-gas emissions). This comparison indicates that human greenhouse-gas emissions, volcanic, and solar effects have all contributed to global temperature changes of the past century, but that greenhouse-gas emissions are responsible for the great majority of the rapid warming seen in the past few decades. This conclusion is supported by direct measurements of solar output and volcanic activity, which show that these two factors have not changed as they would have to, to have caused any substantial fraction of the late-twentieth-century warming.

In view of this compelling combination of evidence from multiple sources, the IPCC has concluded that "... most of the observed warming over [the years 1950–2000] is likely to have been due to the increase in greenhouse gas concentrations." And since we know that human activities are responsible for recent increases in atmospheric concentrations of greenhouse gases over this time, this means that humans are responsible for most of the rapid warming of the past 50 years.

It is also important to note how this conclusion is limited. The Earth has been warming for a century or two, perhaps more, depending on which data source you consider. For the warming before 1950, human emissions probably played a role but it is likely that other factors, such as a brightening Sun, also contributed. But for the rapid warming of the last few decades of the twentieth century, human emissions can account both for the magnitude of the warming and for various details of its timing and distribution, while no other proposed cause can account for more than a small fraction of it. Consequently, we can conclude with high confidence that human greenhouse-gas emissions are the dominant cause of this rapid recent warming.

3.3   What future changes can we expect? Predicting climate change over the twentyfirst century

While determining how and why climate has changed in the past is important, it is the threat of future climate changes that drives public concern and policy-making. Making informed decisions of what to do about climate change requires information about what climate changes we might face in the future, and how our actions can moderate them. This need puts predictions of future climate change at the very heart of the policy debate.

The primary tool for predicting future climate is the climate model. In the previous section, we showed how climate models help us attribute the Earth's late-twentieth-century temperature rise to human greenhouse-gas emissions. Those climate simulations used actual, measured atmospheric concentrations of CO_2 and other greenhouse gases as inputs from which to simulate the historical climate.

Model predictions of future climate change also require atmospheric concentrations of CO_2 and other greenhouse gases as inputs, but in this case these values cannot be measured but must be predicted. Predicting future atmospheric concentrations requires predicting how much CO_2 and other greenhouse gases human activities will emit. Emissions predictions, however, are not a matter of atmospheric science, but an exercise in predicting demographic, economic, technological, and social trends over a period of a century or more.

Future emissions will depend on several factors. First, they will depend on global population trends, because emissions are generated by energy use, industry, agriculture, and other activities that usually increase with the size of the population. Second, emissions will depend on world economic growth, because as people grow more affluent they generally demand more energy-consuming goods and services. Third, emissions will depend on technological trends that determine the efficiency of energy use and the mix of carbon-emitting and non-carbon-emitting energy sources in the economy. Emissions will also depend on policies, whether these are undertaken for other purposes (for example, policies to promote economic growth or technological innovation, to influence energy supply or demand, or to control population growth, such as China's one-child policy). And finally, they will depend on large-scale historical events that we have little ability to predict or control, such as major wars, political transitions, or the emergence of epidemic diseases.

We do have some knowledge on which to base projections of these demographic, economic, technological, and political trends. Historical experience provides guidance regarding ranges of trends in population and economic growth and technological innovation that are likely. We can also exclude some combinations of trends as highly unlikely. For example, since technological innovation in an economy tends to track new investment and therefore economic growth, it is most unlikely to have an extended period with both stagnant economic growth and rapid technological innovation.

But we do not have enough knowledge of the processes shaping future emissions to make a single prediction to which we can grant much confidence. Indeed, there are large uncertainties and frequent errors even in one-year predictions of economic growth. Rather, our knowledge admits a fairly wide range of possible futures, which we can represent by a set of "scenarios" of future emissions. Each scenario provides an alternative, internally consistent, plausible picture of how world development might shape emission trends over the twentyfirst century. Together, the set of scenarios should span the range of alternative emission futures that we judge plausible.

As part of the IPCC's job of summarizing scientific knowledge of climate change, IPCC working groups have conducted two major exercises to develop scenarios of greenhouse-gas emissions through the twentyfirst century. The purpose of these scenarios has been to provide plausible, consistent emission inputs to drive climate-model projections. The IPCC's first major scenario exercise, completed in 1992, developed five alternative scenarios. Most subsequent analyses focused on the middle or "reference-case" scenario of these five (named "IS92a"), which projected that world greenhouse-gas emissions would grow from their present 8 GtC/yr,[5] to about 14 GtC/yr by 2050 and 20 GtC/yr by 2100.[6]

The IPCC's recent, more detailed scenario-development exercise avoided defining a central case, but rather identified four alternative families of scenarios (called A1, A2, B1, and B2), each of which presented a distinct pattern of world development. From a total of 35 scenarios in the four families, six "Marker Scenarios" were selected to serve as benchmarks for climate model projections in the IPCC's third assessment report (2001a): one scenario from each family, plus two

---

[5] "GtC/yr" means "gigatonnes of carbon per year." A gigatonne is a billion metric tons (also called tonnes), and a metric ton is 1000 kg, or about 2200 lbs.

[6] The units are gigatonnes of carbon-equivalent. This means that projected emissions of non-CO_2 gases are converted to an equivalent quantity of CO_2 that has the same heat-trapping effect. This total quantity of CO_2-equivalent is then measured by the mass of the carbon within it (i.e. excluding the mass of the oxygen atoms in the CO_2 molecule).

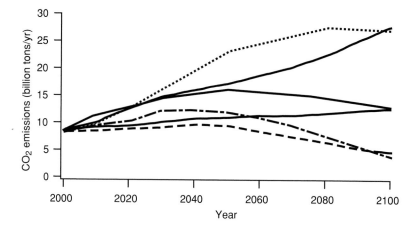

Figure 3.11. CO_2 emission scenarios used in models to predict future climate change. The dotted line is the scenario A1FI, the dashed line is the scenario B1, the dot-dashed line is scenario A1T. See text for a short description of these scenarios. Source: Fig. 17 of the Technical Summary, IPCC (2001a).

technological variants from the A1 family. Figure 3.11 shows projected greenhouse-gas emissions for these six marker scenarios through 2100.

Projected CO_2 emissions in these scenarios span a wide range. Starting from present emissions of about 8 GtC/yr, projected emissions in 2100 can be as high as 30 GtC/yr or as low as 5 GtC/yr. This huge disparity of possible emission futures reflects the combinations of uncertainties about population, economic growth, and technological trends. For example, the dotted line, which shows the largest emission growth over the century (called A1FI in the IPCC scenario exercise), assumes a moderately optimistic continuation of recent trends – relatively low population growth, high economic growth, and gradual convergence of incomes between world regions – but also assumes that fossil fuels remain the primary source of most of world energy, with a shift toward coal as lower-cost production of oil and natural gas declines. Under this scenario, projected emissions reach 30 GtC/yr by 2100. The dashed line, which shows the lowest cumulative emissions, represents a highly optimistic version of what might be called a "sustainable devel-opment" future. It projects the same population growth as the dotted line, some-what slower economic growth but with lower energy and material intensity due to a shift of the world economy from manufacturing to services and information, as well as rapid adoption of cleaner energy technologies. Under this scenario, called B1, emissions peak below 10 GtC/yr around mid-century, then decline to about 5 GtC/yr, below present levels, by 2100.

Also note the dot-dashed line. This scenario, called A1T, assumes the same patterns of population and economic growth as the dotted line, A1FI, but starkly different technological trends. Instead of increasing dependence on higher-carbon fossil fuels like coal, this scenario assumes technological development which allows a shift toward non-fossil (or non-CO_2 emitting) energy sources as low-cost oil and gas decline. Emissions under this scenario rise through mid-century to between 10 and 15 GtC/yr, then decline to even lower than those under the B1 scenario. This comparison shows the central importance of technological development for emissions and climate futures. Chapter 4 provides more detail about the assumptions underlying these scenarios and what they imply for possible actions.

An important additional area of uncertainty in climate-model projections concerns how they treat aerosols – tiny particles, either solid or liquid, suspended in the atmosphere. Fuel combustion and other human activities release various types of aerosols to the atmosphere, which can either warm or cool the Earth's surface depending on their composition. Black carbon aerosols (tiny particles of soot) absorb both incoming solar and upwelling infrared radiation, and so warm the surface. Liquid sulfate aerosols reflect incoming solar radiation back to space, and so cool the surface. Because we do not have good understanding of either the present global distribution of the different types of aerosols, or of how they are likely to change in the future, they create uncertainties for both understanding present atmospheric conditions and projecting future trends. If emissions of black carbon aerosols decrease strongly in the future, this will tend to cool the surface and offset some projected warming from CO_2 and other greenhouse gases; if they increase, this will increase projected warming. The opposite is true for sulfate aerosols: if they decrease, this will add to projected warming; if they increase, this will introduce a cooling effect that partly offsets projected warming. Although much research is underway on the distribution, sources, and sinks of various types of aerosols, they still represent one of the largest sources of uncertainty in present climate-model projections.

A particular climate model, using a particular emissions scenario as the input, will generate a projection of climate change over the next century. But there are a dozen or so current-generation climate models in use, each developed by a different scientific group. These models differ in their approaches to simulating the atmosphere. They may break the atmosphere up into different size boxes, give greater or lesser emphasis to certain processes in the atmosphere, or use different computational approaches to represent basic climatic processes, particularly those that operate at scales finer than the smallest cells in the model and therefore cannot be represented explicitly. Because of these differences, different models project different climate futures, even when driven by the same scenario of future emissions. A standard yardstick to compare climate models is their "sensitivity," defined by how much eventual warming they project when the pre-industrial atmospheric concentration of CO_2 is suddenly doubled (from about 270 to

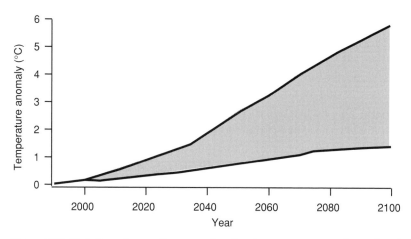

Figure 3.12. Globally and annually averaged surface temperature anomaly predicted by climate models for the twentyfirst century, measured relative to 1990. Source: Figure 5 of the Summary for Policymakers IPCC (2001a).

550 parts per million by volume), then held at that higher level forever.[7] The accepted range of climate sensitivity among current climate models is 1.5–4.5 °C. This range, which has remained unchanged for about 25 years, can be considered a measure of the uncertainty in the models.

This range of responses to doubled CO_2 is not the same as a projection of how fast the Earth will actually warm, however. Projecting future warming requires considering both uncertainty in future emission trends and uncertainty in the climate's response to emissions. Figure 3.12 provides this projection for the twentyfirst century. It shows the range of globally averaged warming projected by several climate models that span the accepted range of climate sensitivity, driven by all 35 IPCC emissions scenarios from the four scenario families. By 2100, the range of projected annual-average warming is 1.4–5.8 °C above the 1990 level.[8]

This is a sobering picture. While there is a wide range of uncertainty in the magnitude of future warming, warming is projected to continue through the twentyfirst century under all models and all emissions scenarios. Moreover, even the bottom of the range, an increase of 1.4 °C by 2100, is double the 0.6 °C warming

[7] In the eighteenth century, before the industrial revolution, atmospheric CO_2 abundance was about 270 parts per million (p.p.m.). Today's atmospheric CO_2 abundance is about 380 p.p.m. This means that CO_2 makes up 0.038% of the atmosphere's volume.

[8] Because it is impractical to do many experimental runs with full climate models, this range was actually generated by calibrating a simple climate model (without fine-scale detail) to seven complex models, with sensitivities from 1.7 to 4.2 °C and a range of response times. Projected warming is greater than in the 1995 IPCC report principally due to an updated and more realistic projection of how fast sulfur emissions will decline with increasing control of conventional air pollution and acid rain.

of the twentieth century. This would represent a rate of change similar to the rapid warming experienced since 1975, but continued for 100 years. If future warming falls in the middle of the range in Figure 3.12, which we must assume is a more likely outcome, or near the top of the range, then the rate of warming over the century would be extreme. The only historical evidence of global temperature changes as large and as fast as the high projections for this century is a series of abrupt warmings and coolings that occurred at the end of the last ice age. But these changes accompanied major reorganizations of the circulation of the atmosphere and ocean, which are not now occurring. Consequently, the upper range of projected warming this century may represent a climate change that has no precedent over the entire history of the Earth.

To summarize, despite uncertainty in both emission projections and climate models, there is substantial scientific agreement that it is *nearly certain* that the Earth's temperature will continue to increase, and it is *highly likely* that by the end of the twentyfirst century, the global average temperature will be several degrees Celsius warmer than the present.

3.4　What will the impacts of climate change be?

Changes in annual and globally averaged temperature are the yardstick we use to describe the magnitude of climate change. But while these projected global temperature changes look extremely serious, few people care about global-average temperature for its own sake. Rather, climate change matters because of its potential impacts on things people value, such as freshwater availability, food production, human health, recreational opportunities, risks of extreme weather such as severe storms, floods, or droughts, and the aesthetics of the outdoors.

Predictions of the impacts of climate change are more difficult and uncertain than predictions of changes in the globally averaged climate, for several reasons. First, no one lives in the global-average, annual-average climate, so understanding climate impacts requires projections at the finer scales, both in space and in time, where people and climate-sensitive systems actually experience the climate. This requires projections region by region, since the effects of a few degrees' warming over a desert and over a forest would be completely different. It also requires projections by season, since the effects of a few degrees' warming in summer and in winter would also be completely different. Such small-scale and seasonal predictions are far more uncertain than model predictions of the globally and annually averaged climate. Second, impacts may depend on many dimensions of climate change – for example on precipitation, humidity, or winds in addition to temperature, and on changes in variability and extremes in addition to changes in average values – that are also more difficult to project and more uncertain than changes in average temperature. Third, projecting impacts requires estimating the responses of climate-sensitive ecosystems, resources, or activities to projected climate change. Adding this extra step (for example, how does the forest respond to a specified climate change) further increases the uncertainty in projections. A final difficulty in projecting impacts is that many areas of likely climate impact, such as agriculture and commercial forests, are dominated by human management. In these areas, little can be said about climate impacts without considering human responses, including measures people may take to adapt to climate change. We discuss socio-economic aspects of impacts and potential adaptation measures in the next chapter. In this section we summarize present knowledge about impacts, focusing mainly on those direct bio-physical changes with the best prospects for understanding them separately from human responses.

Despite these difficulties, we do know some things about the likely impacts of climate change, and are making progress in understanding both the likely direction and in some cases the approximate magnitude of many impacts. Our knowledge is highly variable across different types of impact, however: some are quite well understood, others very weakly; and no doubt, there are many complex connections that may magnify or dampen impacts that we have not even recognized. We know that some impacts are harmful, others are beneficial, and many are mixed – harming some people, places, or activities, and benefiting others. Most analyses of impacts suggest that harmful impacts are likely to outweigh beneficial ones, by a little in places that are rich, well governed, and adaptable, and by a lot in places that are less fortunate. If climate changes are large or happen quickly, harmful impacts are increasingly likely to dominate beneficial ones even in rich and well-governed regions. Nevertheless, predictions of exactly what will happen, where, and when, however, must still be considered highly uncertain.

In some cases where our understanding is good, it is because impacts at regional or even local scale are strongly connected to global-scale climate changes by well-known physical processes. A particularly clear example is sea level rise. A warming climate will continue to raise the sea level, through thermal expansion of seawater and melting of glaciers. The projected range of global-average temperature increase shown in Figure 3.12 is estimated to translate into a rise in sea level of 10–90 cm during the twentyfirst century. Since a half-meter rise in global sea level, roughly speaking, means a half-meter rise along every coastline, what this change will mean for any specific location can be readily, if approximately, assessed. How serious it will be in any particular place will depend on local factors, for example the amount of low-lying coastal land, whether the land is rising or sinking locally, the pattern of settlement, use of and investment in the land, and the resources available to manage an appropriate mix of coastal protection and orderly retreat.

Potential changes in tropical storms are another regional impact strongly linked to global-scale processes. Because the strength of tropical cyclones (hurricanes and typhoons) depends on sea-surface temperature in tropical latitudes, there is good basis for expecting that the maximum intensity of these storms will increase as the climate warms. Whether the intensity of mid-latitude storms will also increase is more uncertain. These storms do not have as specific a linkage to a global temperature trend, and some climate-model projections show them increasing in intensity while others do not.

Other impacts require examining climate projections at regional scales. As mentioned above, climate-model projections grow more uncertain as you consider smaller regions. But some broad regional results are now well established, because they appear consistently across many climate-model projections and are grounded in basic physical principles. For example, temperatures are likely to increase more over land than over sea, because of the moderating effect of the ocean's huge heat capacity. In most locations more warming is projected at night than in the day and more in winter than in summer, reducing both daily and annual temperature ranges. More warming is projected at middle and high latitudes, particularly in the Northern Hemisphere, than in the tropics. For example, recent climate model simulations project warming in northern North America and Eurasia more than 40 percent greater than the global-average warming.

The Arctic and sub-Arctic regions will experience extreme warming, with severe implications for many resources and human activities. The impacts of climate change can already be observed here more clearly than anywhere else, due to the sharp warming of the past few decades. Thawing of permafrost, retreat and thinning of sea ice with resultant increases in coastal erosion and disruption of marine ecosystems, and shorter ice-travel seasons on lakes and rivers, have already brought disruptive impacts to Arctic regions. Since ice reliably melts at 0 °C, these impacts are likely to accelerate under the large warming projected for the twentyfirst century: 4.0–7.5 °C over Arctic land areas in summer by 2080, 2.5–14 °C in winter. All climate models project continued retreat of Arctic sea ice through the twentyfirst century, and some of them project completely ice-free summers in the Arctic Ocean by the end of the century. An even partially navigable Arctic Ocean would have huge effects on shipping, Arctic development, and military operations and security. In addition, the substantial loss of summer Arctic sea ice would have implications for global ocean circulation that are not yet well understood, but potentially enormous.

Continental regions in the middle latitudes are projected to experience warming (about 3–6 °C over the continental United States, for example) although projections of how this average warming will be distributed across the continent are variable. The combination of warmer summer temperature and increased humidity is likely to bring substantial increases in the summer heat index,[9] with some models predicting an increase as large as 5–14 °C in July in the southeastern states of the USA. If you live there, you know how miserable that would be.

Average precipitation is also projected to increase, although there is more uncertainty about where the largest precipitation changes will occur than in the corresponding projections for temperature. Continuing a trend of the twentieth century, more of the total annual rainfall is projected to come in the heaviest downpours, bringing increased erosion and higher risk of flooding and landslides. Moreover, when rain falls in heavy downpours, more of it runs off and less is absorbed by soil or stored in reservoirs for human use. Combined with warmer summers, which will increase the rate at which water is lost from soils by evaporation, this leads to the surprising result that both wet and dry extremes will grow more likely: wet extremes, with associated risks of flooding, increased erosion, and landslide; and dry extremes, with associated risks of water shortages, crop loss, wildfire, and increased vulnerability of crops and forests to pests and disease.

The significance of these climate impacts to human affairs is obvious. But assessing other impacts requires detailed analysis of the behavior of climate-sensitive systems. For example, rainfall is the major source of the freshwater that human and natural systems depend on. Changes in the amount, location, and timing of precipitation can therefore alter freshwater availability. However, quantitative projections of freshwater changes require a detailed understanding of specific water systems, as well as how people manage them. A study of climate change effects on the Columbia River Basin in the US Pacific Northwest provides an important recent example of the kind of analysis required. While projected changes in total annual precipitation and streamflow through the twentyfirst century in the Columbia River Basin are small, the warmer wetter winters and hotter drier summers lead to a sharp shift in the seasonal pattern of streamflow. Because the Columbia River Basin draws much of its flow from the melting of accumulated winter snowpack, its flow presently peaks in the late spring. But under projected warmer winters, much more of total annual precipitation will fall as rain rather than snow. This will increase the river's flow in winter, when water is already abundant in the region, and decrease it in summer, when the region is already acutely water-scarce. Similar changes are likely in other regions that meet water needs in dry summers by drawing on snow-fed river systems, highlighting the importance of examining not just annual total water availability, but details of seasonal flows.

---

[9] The heat index combines temperature and humidity to produce a measure of how hot it feels like. This is similar to the concept of wind chill, which combines temperature and wind speed in cold conditions to create a measure of how cold it feels.

Climate change will also affect natural or unmanaged ecosystems. The distribution of plant, animal, and microorganism species are influenced by many factors, but climate is a major determinant. Changed climate will affect many aspects of the reproduction, behavior, and viability of species in diverse ways, and consequently their spatial range, as well as relationships among species. There is abundant evidence that these changes are already underway in response to recent climate change, including shifts of species ranges poleward and to higher elevations, and changes in the timing of seasonal events such as tree leafing, leaf-fall, and egg-laying. Under continuing climate change, present ecosystems will not simply move intact to follow the optimal climate: each species will be affected in particular ways, and ranges will adjust at different rates and by different processes, in many cases subject to other human interventions and constraints such as land-use change, barriers, and intentional or inadvertent transport.

The aggregate result will be that present ecosystems are continually disrupted and reorganized, with new relationships among incumbents and new arrivals continually re-established in each location. In some cases, the new assemblies may be similar enough to present systems that thinking of present ecosystems simply being shifted (for example mixed-temperate forests shifting north into the present boreal forest zone, boreal forests shifting north into the present tundra zone) is not too misleading. In other cases, however, the new systems may be unlike present ecosystems, with far-reaching effects for ecosystem services such as water retention and nutrient cycling, and for ecosystem amenities such as opportunities for human uses and recreation. A few major ecosystem types are likely to be lost entirely, because of physical limits or barriers to the movement of key species, or the complete loss of the required climate conditions. In the United States, ecosystem types threatened with total or near-total loss include alpine systems in the lower 48 states, coastal mangrove systems, coral reefs, and arid ecosystems in the southwestern states. A particularly important factor in ecosystem impacts will be the rate of climate change. Ecosystems have adapted to climatic variations in the past, but the changes have been much slower than those projected for the coming century. It is virtually certain that ecosystems will adapt less gracefully to the predicted rapid changes than to the slow changes of the past few thousand years: what is highly uncertain is how much less gracefully.

Systems managed for human use, such as agriculture, commercial forests, rangelands, and aquatic and marine systems (fisheries, etc.), are also sensitive to climate and related changes, but are dominated by human management decisions. In terms of projecting climate-change impacts, this dominant effect of human management cuts two ways. On the one hand, serious disruption of these systems from climate and related changes may have severe human impacts because we depend on them so much. On the other hand, the ability to adapt management practices to changing conditions offers the possibility of mitigating these harmful impacts. We discuss the linked issues of impacts and adaptation to climate change in the context of other forms of change in the next chapter.

We can roughly summarize present knowledge about the impacts of climate change as follows. For the rich, mid-latitude countries (for example, the USA, Europe, Japan), impacts of climate change might range from small to severe over the twentyfirst cetury. These countries, however, have substantial financial, technological, managerial, and political capacity to adapt to harmful impacts – unless climate change lies near the top of the projected range shown in Figure 3.12, in which case even these countries are likely to face serious challenges. Poorer countries, mostly located in the tropics and sub-tropics, are projected to face climate changes different in detail but at least as challenging as those projected for the mid-latitudes. Because these countries have fewer resources to adapt to impacts, the consequences for them may be severe even for climate change in the middle of the projected range.

Moreover, it is crucial to note that none of the projected changes will stop in 2100. Stabilizing atmospheric concentrations will require that emissions eventually decline to near zero, but most scenarios show continued growth of emissions beyond 2100. Uncertainties in climate projections continue to grow larger beyond that point, of course, but the few analyses that have looked further suggest that climate change and its impacts grow increasingly more severe. There is growing evidence that benefits to plant growth from elevated CO_2 level off over time as CO_2 continues to rise, while stresses from climate change continue to increase. Consequently, unless there is an extreme level of technological and economic progress that frees us from dependence on anything resembling crops in fields or relatively natural forests – which may well happen, since 100 years can bring vast economic and technological changes – impacts beyond 2100 look serious and potentially unmanageable, even for the rich countries of the world.

Even if atmospheric greenhouse-gas concentrations are stabilized, the climate will continue to change and its impacts will continue to compound because of lags in the climate system. The oceans take thousands of years to warm up, so additional climate change and impacts will continue to accumulate for a millennium or so, even after atmospheric CO_2 is stabilized. For example, if atmospheric CO_2 were stabilized by around 2100 (by which time sea level would have risen 10–90 cm), sea level would still rise a further 1–2 meters over the next few centuries, due to continued thermal expansion and glacier melt.

Finally, it is also necessary to consider the possibility of climate surprises: high-consequence, possibly sudden changes that either appear to be quite unlikely (but which cannot be ruled out), or that we may completely fail to anticipate. An example of such a potential extreme event would be the disintegration of a major continental glacier. The West Antarctic Ice Sheet contains several million billion tons of water. We know that this ice sheet has disintegrated in the past, raising sea level by 4–6 meters (about 13–20 feet). Most experts think that for it to collapse again would require warming of 8–10 °C and take more than a century, but there are some (controversial) indications that it might be possible with less warming (as little as 4 °C in the relevant region) and could occur much faster. There are also recent – and controversial – suggestions that the Greenland Ice Sheet could be vulnerable to melting over a century or so from as little as 1 to 3 °C warming, raising global sea level by about 7 meters (about 24 feet). The flooding of coastal areas worldwide that would result from either of these events would represent an unimaginable environmental and humanitarian catastrophe. Most experts think these events are unlikely to happen in the next few centuries, but the risk of them happening much faster cannot be ignored.

3.5   Conclusions

We conclude by summarizing the answers that present scientific knowledge provides to the four key questions about global climate change.

Is the Earth's climate getting warmer? Definitely yes. Multiple independent data sources confirm beyond any reasonable doubt that the Earth's surface warmed during the twentieth century, with particularly rapid warming over the last few decades. There is some indication that the warming has been going on for several centuries before the twentieth century, but the data are less clear on that.

Are human activities responsible for the observed warming? Probably. It is likely that greenhouse-gas emissions from human activities have caused most of the rapid warming in the last few decades of the twentieth century. The warming that occurred before about 1950 was probably caused by a combination of factors. Emissions from human activities probably played some role, but natural processes such as solar variability, volcanoes, and internal climate variability probably also made substantial contributions.

What future climate changes can we expect? Although projecting the precise magnitude and regional details of future climate change involves much uncertainty, the pressure exerted by emissions from human activities is already substantially altering the Earth's climate. Consequently, as the scale of human activities grows further over the twentyfirst century, it is virtually certain that the Earth will continue to warm, and highly likely that the total global warming over the century will be between 1.4 and 5.8 °C. Even the bottom end of this range would be more than double the warming of the twentieth century.

What will the impacts of future climate change be? We have a broad idea of the types of regional changes and impacts that are likely, but cannot predict specific impacts with confidence. The range of possible future impacts includes some that are serious enough to compel our attention. If climate change lies near the low end of the projected range, impacts over the twentyfirst century are likely to be manageable for rich, mid-latitude countries, but may pose serious difficulties for poorer countries. If climate change lies near the high end of the projected range, impacts over the twentyfirst century are likely to be severe and potentially unmanageable for everyone. Continued changes after 2100, with atmospheric CO_2 increasing beyond triple or quadruple the pre-industrial level, are even more uncertain but include a non-negligible risk of severe impacts, including abrupt climate changes, that would represent grave threats to rich and poor countries alike.

Further reading for Chapter 3

IPCC (2001a). *Climate Change 2001: The Scientific Basis. Contribution of Working Group I to the Third Assessment Report of the Intergovernmental Panel on Climate Change*, ed. J. T. Houghton, Y. Ding, D. J. Griggs, M. Noguer, P. J. van der Linden, X. Dai, K. Maskell and C. A. Johnson. Cambridge and New York: Cambridge University Press.

> This is the most recent full-scale report of the IPCC's Working Group I, the group responsible for international assessments of the atmospheric science of climate change. It is the most recent authoritative statement of the status of scientific knowledge about climate change, and an essential source for anyone wishing to be literate in the climate change debate. In addition to the fully detailed and cited syntheses of specific aspects of climate-change science presented in each chapter, the report includes a technical summary and a policy-makers' summary that present the most important results and conclusions in more condensed and accessible form. We draw extensively on this report for many of the scientific conclusions we present in this chapter.

IPCC (2001b). *Climate Change 2001: Impacts, Adaptation, and Vulnerability. Contribution of Working Group II to the Third Assessment Report of the Intergovernmental Panel on Climate Change*, ed. J. J. McCarthy, O. F. Canziani, N. A. Leary, D. J. Dokken and K. S. White. Cambridge and New York: Cambridge University Press.

> This is the most recent full assessment of the IPCC's Working Group II, which summarizes present knowledge about potential impacts of climate change, ability to adapt, and vulnerability of environmental and social systems to climate change.

IPCC (2001d). *Climate Change 2001: Synthesis Report. A Contribution of Working Groups I, II, and III to the Third Assessment Report of the Intergovernmental Panel on Climate Change*, ed. R. T. Watson and the Core Writing Team. Cambridge and New York: Cambridge University Press.

> This report summarizes and integrates the principal results of the three IPCC working groups into a single volume.

IPCC (2000). *Emissions Scenarios: Special report of the Intergovernmental Panel on Climate Change*, ed. N. Nakicenovic and R. Swart. Cambridge and New York: Cambridge University Press.

This special report provides the results and background for the IPCC's most recent set of emission scenarios, which were used as inputs to the climate-model projections in the Working Group 1 report and are summarized in this chapter.

Karl, T. R. and Trenberth, K. E. (2003). Modern global climate change. *Science*, **302** (5 December), 1719–1723.

A brief, moderately technical summary of current knowledge of global climate change and human influences on it, including several advances too recent to have been reflected in the 2001 IPCC report.

US Global Change Research Program, National Assessment Synthesis Team (2001). *Climate Change Impacts on the United States: The Potential Consequences of Climate Variability and Change*. New York: Cambridge University Press.

This report, produced by the US Global Change Research Program, provides a more detailed assessment of potential climate-change impacts, vulnerabilities, and capacity for adaptation for the United States. Separate studies examine effects of climate change on nine major US regions and five sectors of national importance. Like the IPCC reports, this assessment involved the work of hundreds of scientists and was subjected to a rigorous and thoroughly documented process of peer review. It is often referred to as "the US national assessment."

United States National Academy of Sciences, Climate Research Committee, Panel on Climate Change Feedbacks (2003). *Understanding Climate Change Feedbacks*. Washington, D.C.: National Academy Press.

This report looks at what is known and not known about climate change feedbacks and seeks to identify the climate feedback processes most in need of improved understanding. The report suggests an approach by which progress toward better understanding of climate feedback processes can be made.

United States National Academy of Sciences, Commission on Geosciences, Environment and Resources (2000). *Reconciling Observations of Global Temperature Change*. Washington, D.C.: National Academy Press.

This report discusses the technical intricacies of the surface thermometer, balloon, and satellite records. It also discusses the uncertainties in each data set, technical issues in comparing trends derived from the different data sets, and possible explanations for the disagreement between these sources.

Weart, S. R. (2003). *The Discovery of Global Warming*. Cambridge, MA: Harvard University Press.

A highly readable and accessible history of major developments in the science of climate change, from the nineteenth century through the formation of the modern consensus about the reality and predominant human cause of recent climate change as expressed in the 2001 IPCC report.

4

The climate-change policy debate: impacts and potential responses

An understanding of the science of climate change provides only part of what is needed to decide what to do about the issue. We also need information about the likely impacts of climate change on human society, the options for responding to climate change, and the relevant tradeoffs among policy choices with their associated effectiveness, benefits, risks, costs. This chapter summarizes present knowledge and uncertainties on these matters.

The responses available to deal with the threat of climate change can be grouped into three broad categories. **Adaptation** measures target the impacts of climate change, seeking to adjust human society to the changing climate and so reduce the resultant harms. Building seawalls would be one way to adapt to sea level rise; planting drought-resistant crops would be one way to adapt to drier agricultural regions. **Mitigation** measures – an odd use of the term, but one too well established in the policy debate to resist – target the causes of climate change, seeking to reduce the emissions of greenhouse gases that are causing the climate to change.

Most proposals to address climate change revolve around mitigation and adaptation. A third class of potential responses involves actively manipulating the climate system to offset the climatic effects of greenhouse-gas emissions, making it possible to break the linkage between emissions and climate change. This approach, sometimes called **geoengineering**, has received less attention than mitigation and adaptation, and present understanding of its potential benefits and associated costs and risks is in its infancy. Still, we will argue that this approach also merits serious examination, particularly if the severity of climate change turns out to lie near the upper end of current projections.

Section 4.1 discusses impacts of climate change and adaptation measures. Section 4.2 discusses projections of emissions over the next century and mitigation strategies to reduce them. Section 4.3 discusses the attempts that have been made to assess the costs of climate impacts, adaptation, and mitigation measures, and to integrate these into a consistent framework to decide how to respond. Section 4.4 presents a brief discussion of geoengineering measures, while Section 4.5 presents brief conclusions about the problem of making decisions about how to respond to climate change under uncertainty.

4.1   Impacts and adaptation

4.1.1   Defining and assessing the impacts of climate change

Chapter 3 discussed present knowledge of how the climate is likely to change over the next century, and the resultant impacts on ecosystems and natural resources – projected changes in temperature, precipitation, agricultural crop yields, the range and composition of specific forest types, seasonal streamflows in major river systems, etc. Determining the impact of these changes on people requires additional analysis, because these projections of climate change and its direct impacts on ecosystems and resources must be integrated with information about the societies on which the changes are imposed. The effects of any specified climate change on particular people or communities will depend on a host of socio-economic details, such as where and how people live, how rich or poor they are, how they earn their livings, what technologies and natural resources they rely on, and what policies and institutions govern them. Consequently, impacts will vary among people and places not just because of differences in how the climate changes but because of differences in the factors that make people more or less vulnerable to particular changes. For example, an electric utility might be sensitive to the frequency and length of summer heat waves, which raise the demand for electricity, while a ski resort might be highly sensitive to changes in average winter temperature and total snowfall but not at all to summer temperatures. Agriculture in a particular region might be sensitive to changes in total growing-season precipitation and the frequency of heavy downpours and droughts. Low-lying coastal areas, whether coastal Louisiana or Bangladesh, may be especially vulnerable to sea level rise, while rapidly growing regions with freshwater scarcity such as California are vulnerable to changes in annual precipitation or winter snowpack. Projecting impacts of climate change requires constructing detailed scenarios of future society with plausible assumptions about the socio-economic factors that are the most important contributors to such vulnerabilities.

In addition, the impacts of climate change will also depend on how narrow or broad a field of view is considered. Impacts on human society are likely to be larger and more diverse, including both gains and losses, when smaller regions and narrower segments of the economy are considered. Even within a single small region, warmer and drier summers might harm agriculture but benefit tourism.

Considering the local economy as a whole, the harms to farmers and benefits to tourism would partly offset each other, giving smaller overall impacts. The more broadly you look, the more such smaller-scale effects will be averaged out. Consequently, any national or global projection of climate impacts will combine and conceal great variation in smaller-scale impacts, by which some people, places, and activities may be harmed, while others – at least for small changes in climate – may benefit.

A basic challenge in assessing climate-change impacts is projecting how well people and organizations will adapt to the changes. We know that human society is adapted to present climates in diverse ways, and we expect some adaptation to future changes. If climate change reduces the yields and profits from present farming practices, we expect farmers – and all the others whose choices influence agricultural practices, such as seed and equipment companies, agricultural extension services, and researchers – to shift to crops and practices better suited to the new conditions. If present settlement patterns, economic activities, or management of water, forests, or other natural resources come to be ill-suited to a changed climate, we expect people to notice and change their practices to better match the new climate – eventually, and to some degree. Moreover, people need not necessarily wait for a change to happen to adapt to it. If good forecasts of likely future climate changes are available, people may look ahead and adapt in advance, either to the specific changes they expect, or to a general increase in uncertainty about future climate. Such anticipatory adaptation is especially important for decisions whose consequences extend many decades into the future, such as zoning and settlement policies, and long-lived capital investments such as ports, bridges, dams, and power plants.

The impacts actually experienced from future climate change will depend strongly on adaptation, so it is essential to consider adaptation in attempting to assess impacts. Unfortunately, we know little about how well and how quickly people adapt, or about the factors that promote or constrain their capacity to adapt. Most assessments of climate-change impacts have made one of two extreme assumptions about adaptation. One extreme assumes that present practices continue unchanged, with no regard for the changed climate conditions. By excluding any adaptation, this approach systematically overstates harms from climate change. The opposite extreme assumes optimal adaptation, unconstrained by such limits as imperfect foresight or rigidities due to long-lived capital equipment. Just as assuming no adaptation predictably overstates harms, assuming ideal adaptation understates them. Indeed, by assuming that adaptation to future changed climatic conditions will be substantially *better* than present adaptation to present climate conditions, it is possible to project, implausibly, that the impacts of nearly any climate change will on balance be beneficial.

Better impact projections will depend on realistic estimates of how well and how fast people will actually adapt to future climate changes, but this is a wide open question. The capacity for adaptation varies strongly among people and places. Rich societies with well functioning institutions are more able to adapt, and consequently less vulnerable to climate change, than those without such advantages. Having the capacity to adapt, however, does not necessarily mean that effective adaptation will take place. There is ample evidence that we are not ideally adapted to the present climate. For example, we operate intensive agriculture in drought-prone regions, dependent on the unsustainable mining of groundwater. We build in high-risk locations on low-lying coastlines, flood-prone river valleys, and fire- and slide-prone hillsides, and even rebuild repeatedly in the same locations, often with public subsidies, after property is destroyed. Such maladaptations leave us more vulnerable than necessary both to present climate variability and to projected future changes. Perhaps we will pay more attention to these issues as climate change continues and impacts become more conspicuous, and so will adapt better to future changes than we have to present conditions. Perhaps not.

An additional basic challenge in assessing climate impacts is that climate does not matter in isolation from other environmental changes. As the climate changes, other aspects of the environment will change in parallel: atmospheric CO_2 will certainly increase, and other changes such as nutrient deposition, air quality, and land-cover are also highly likely. Many human and biological systems are likely to be sensitive to both climate change and these other changes, and to interactions between them. The only such interaction that has been studied extensively is that between climate and atmospheric CO_2. In addition to its effect on climate, elevated CO_2 affects plants directly by increasing the efficiency of photosynthesis and water use, although the nature and size of the effects vary widely among plant species. Studies of many agricultural crops and young trees of a few commercially important species have generally found that this CO_2-fertilization effect can offset the stress of projected warmer, drier summers to yield net increases in growth, at least for the modest changes in average climate projected over the next few decades. Present studies have only considered changes in average climate, however, and have not yet considered changes in climate variability or extreme weather events, such as the increased concentration of precipitation in heavy downpours or the increased risk of droughts. Moreover, there has been little assessment of likely changes in weeds, pests, and diseases under climate change and higher CO_2, or of interactions with other forms of environmental change such as nutrient availability and air quality. The few studies done so far of weed–crop–pest interactions under changed CO_2 and climate suggest that these may represent the largest effects on crop productivity and could go in either direction.

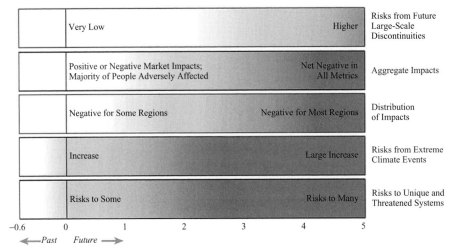

Figure 4.1. A graphical summary of IPCC authors' judgments of how the severity of climate-change impacts are likely to vary with changes in global-average temperature. Each bar denotes a different category of impacts, with the darkest shadings indicating the strongest bases for concern. Source: Figure TS-12, IPCC (2001b). (Note: colored original is converted to gray-scale shadings.)

As this discussion of challenges indicates, specific, quantitative projections of the impacts of future climate change on human society are substantially more difficult and uncertain than the projections of future climate change discussed in Chapter 3, even for specific forms of impact identified in advance such as changes in crop productivity or coastal impacts from rising sea level. Identifying all the potential impacts that will matter and the linkages between them is even harder. But however uncertain impact projections might be, they still must be considered in any reasonable judgment of how to respond to climate change. Waiting for uncertainty in projections of climate change or its impacts to be fully resolved would mean waiting until the changes are actually upon us.

Since waiting this long would involve grave risks, there is great value in further development of methods to assess potential climate-change impacts that consider uncertainty, integrating presently available research and analysis with collective expert judgments where necessary. When such integration has been attempted, the most frequent way has been to roll available knowledge implicitly into some aggregate expert judgment. For example, the authors of the 2001 IPCC report summarized their judgments of the likely severity of aggregate impacts in a chart. This chart, reproduced as Figure 4.1, suggests that severe impacts are likely to mount with average global warming of more than about 3 °C. While such summaries of expert judgment are a useful start, more quantitative and transparent estimates would provide more useful guidance to policy-making.

The need for better methods to integrate uncertainty into climate-change assessments is particularly acute for considering the possibility of abrupt, potentially catastrophic climate changes. As discussed in Chapter 3, a few such mechanisms of abrupt change have been identified as possibilities, such as a rapid disintegration of the West Antarctic ice sheet or a sudden rearrangement of ocean circulation. And there are no doubt other possibilities not yet identified, just as no one anticipated the Antarctic ozone hole until it was observed. We have no basis for thinking that any of these proposed scenarios is likely, and most relevant experts think they are not, but nor do we have any persuasive grounds for saying they cannot happen. It is still not clear how to think about their probability, consequences, or what we might do about them.

4.1.2   Responses to enhance adaptation

People, organizations, and communities will adapt to changing climate conditions on their own, but government policy can also aid adaptation in several ways. Governments might undertake specific adaptation measures by spending public money, for example by building seawalls. Or they might require adaptation measures by private citizens, for example by changing zoning codes to restrict building in coastal areas, flood plains, or other vulnerable areas. Of particular importance is the role of information – climate predictions, impact studies, information about potential responses, or technical assistance – in promoting adaptation. Such information can help people shift from reacting to climate change that has already occurred toward anticipating future changes. Adapting to anticipated climate changes tends to be more effective and less costly, particularly for planning and investment decisions with time horizons of decades or longer. Improved climate analysis and projections can also reduce the risk of private actors mistaking short or medium-term climate variability for a longer-term trend, and so mistakenly adapting to the variability.

Despite expected improvements in climate projections with continuing research, projections will always carry uncertainty. Because the details of the climate impacts we must adapt to will inevitably remain uncertain, adaptation measures will often be directed to making society and its infrastructure robust to a wider range of climate conditions than we presently consider. For example, we might build dams and port facilities higher, build infrastructure of all kinds to handle a wider range of hot and cold or wet and dry conditions, and develop better predictive capability for early extreme storms.

Many potentially valuable adaptation measures will not be specific to climate, but rather will reduce the general vulnerability of people and society to many kinds of risks. For example, strengthening public-health infrastructure will help reduce health risks from climate change, but will also provide enhanced capacity to respond to other health threats. Strengthening emergency-response systems and implementing policies to promote development and poverty reduction would similarly help to reduce vulnerability both to climate change and to other threats.

Measures to promote adaptation will be part of any response to climate change, but it is unlikely that an effective response to the issue will consist only of adaptation. The wide variation across people and societies in capacity to adapt, and the evidence that we are far from ideally adapted to present climate, both suggest limits to how much our total response can or should rely on adaptation. To the extent that adapting to the impacts of climate change is not by itself an adequate response, it is also necessary to target the causes of climate change through mitigation measures, thereby slowing the impacts that we have to adapt to. The next section discusses this approach, and what is presently known about the technical and policy options available to pursue it.

4.2   Emissions and mitigation responses

4.2.1   Emission trends and projections

In Section 3.3 we discussed trends and projections of greenhouse-gas emissions from human activities in the context of predicting how the climate will change over the twentyfirst century. In this section, we provide more detail about the make-up of emissions, the underlying factors shaping their trends and projections, and the means available to reduce them.

The largest source of the human emissions contributing to global climate change is carbon dioxide (CO_2) released by burning fossil fuels: coal, oil, and natural gas. These fuels provide about 80 percent of global human energy use. In 2000, world emissions of CO_2 from burning fossil fuels were about 6.4 billion metric tons of carbon (GtC), an average of about one metric ton per person worldwide.[1] Adding smaller contributions from net deforestation worldwide, principally in the tropics (1.6 GtC), and chemical processes involved in manufacturing cement (0.2 GtC) brought total emissions of CO_2 in 2000 to about 8 GtC.

Several other gases emitted from various industrial and agricultural activities also contribute to climate change. Although emitted in much smaller quantities than CO_2, these other greenhouse gases contribute substantially more warming

---

[1] A metric ton is 1000 kilograms or 2200 pounds, about 10 percent larger than a short ton. A billion metric tons (or a Gigaton) of carbon is often abbreviated 1 GtC. Emissions are sometimes expressed in terms of Teragrams of carbon, or TgC, rather than Gigatons. The relationship between these two units is that 1 GtC = 1000 TgC. By convention, we measure CO_2 emissions by the mass of the carbon contained in them. Occasionally, the mass is reported in terms of the mass of CO_2, which is 3.67 times larger, i.e. 1 ton of carbon is the same as 3.67 tons of CO_2.

per ton emitted, so their total contribution to climate change is similar to that of CO_2. Of these, the most important are methane (CH_4), which is emitted from rice paddies, landfills, livestock, and the extraction and processing of fossil fuels, as well as several natural sources; nitrous oxide (N_2O), which is emitted from nitrogen-based fertilizer and industrial processes as well as several natural sources; and the halocarbons, a group of synthetic industrial chemicals used as refrigerants and in various other industrial applications. Comparing the warming effects of all greenhouse-gas emissions from human activities, methane contributes about one-third as much as CO_2, while nitrous oxide and the halocarbons together contribute another one-third. Emissions of the major greenhouse gases have been growing since the industrial revolution, with the largest increases occurring since the postwar industrial expansion.

Changes in atmospheric ozone, which human activities are increasing in the troposphere (lower atmosphere) and decreasing in the stratosphere (the atmospheric layer above the troposphere, which begins at an altitude of about 10–15 kilometers), probably on balance make a small additional contribution to human-caused warming. Aerosols, small solid or liquid particles suspended in the atmosphere including dust and soot, contribute a mix of less well understood effects, some of which warm and some of which cool the climate.

Although many human activities and several different gases are contributing to climate change, CO_2 from fossil-fuel combustion represents more than half the total contribution. Of present fossil-fuel CO_2 emissions, about 60 percent come from the industrialized countries, 25 percent from the USA alone. The industrialized countries' share of emissions grows larger if you consider cumulative historical emissions (about 70 percent of total fossil-fuel CO_2 emissions since 1950), and smaller if you consider not just fossil-fuel CO_2 but also CO_2 from deforestation and other greenhouse gases (about 40 percent of 2000 emissions).

Turning to projections of future emissions, Section 3.3 (particularly Figure 3.11) introduced the IPCC's six "marker scenarios" of projected emission trends through the twentyfirst century. Under these six scenarios, annual world CO_2 emissions – presently about 8 GtC – ranged from less than 5 GtC to more than 30 GtC in 2100. This large variation among scenarios indicates substantial uncertainty about how emissions will grow, and consequently about how the climate is likely to change, although no specific probability estimates were assigned to the scenarios. (For example, is the probability that emissions turn out to lie within this range 90, 95, or 99 percent? Is the middle of the range more likely than the extremes?)

Despite the wide variation in emissions across scenarios, a few general conditions appear consistently across most or all of them. First, emissions keep growing through the twentyfirst century under nearly all scenarios, as continued world economic growth outpaces emission-reducing technological innovations. Second,

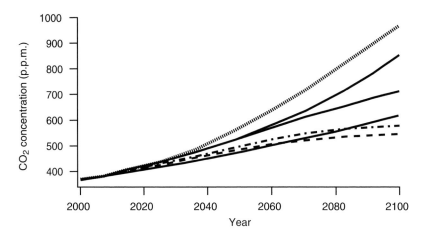

Figure 4.2. Projected abundance of CO_2 in the atmosphere based on the IPCC emission scenarios described in Section 3.3, in parts per million. The dotted line is scenario A1FI, the dashed line is scenario B1, the dot-dashed line is scenario A1T. Source: adapted from Figure 18, IPCC (2001a).

under all scenarios, emissions from the developing countries overtake those of the presently industrialized countries sometime in the first half of the century. Note also that alternative assumptions about trends in energy technology alone can generate the entire range of future emissions, depending on whether the projected decline of cheap oil and gas is followed by a shift toward coal and high-carbon synthetic fuels, or toward sources that emit no CO_2 to the atmosphere.

Figure 4.2 shows the trends in atmospheric CO_2 abundance that follow from these six emission scenarios. All six scenarios generate similar atmospheric CO_2 trends for the next few decades, passing through 500 parts per million (p.p.m.) around mid-century. Only in the second half of the century do different emission scenarios generate large differences in atmospheric concentration. By 2100, the scenario with strong development of non-carbon emitting technology (A1T) has CO_2 nearly stable around 550 p.p.m., roughly double the pre-industrial value, while the scenario with a strong shift to coal-based energy (A1FI) has reached about 900 p.p.m., triple the pre-industrial value, on its way toward a quadrupling in the twentysecond century.

These differences in atmospheric CO_2 will in turn make large differences in climate change, but these will take even longer to be realized due to the inertia of the climate system. For example, one analysis of an aggressive emission-reduction scenario found that in 2050, by which time emissions had been cut by 40 percent, global temperature was only 0.2 °C lower than with no emission reductions.

The other greenhouse gases also contribute to warming and are projected to increase under most scenarios, with the majority of emissions shifting to

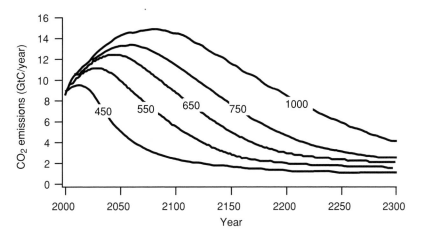

Figure 4.3. CO_2 emission scenarios that would lead to stabilization of the concentration of CO_2 in the atmosphere at 450, 550, 650, 750, and 1000 p.p.m. Source: adapted from Figure 6.1, IPCC (2001d).

developing countries. In contrast to the greenhouse gases, emissions of sulfur dioxide (SO_2) – a pollutant released from burning sulfur-containing fuels, especially coal – cool the climate, because they form aerosol droplets in the atmosphere that reflect incoming sunlight. If SO_2 emissions grow over the next century, they will contribute a regionally concentrated net cooling effect, offsetting part of the warming effect of greenhouse-gas increases; but if they decline, this will make an additional contribution to global climate warming. Although SO_2 emissions have increased over the twentieth century, their growth has slowed and in some regions reversed over the past few decades as they have come under increasingly strict controls to limit acid rain and other environmental harms. Most scenarios project that SO_2 will decline over the twentyfirst century, but some show large near-term growth over the next few decades in the developing countries.

These scenarios all represent informed guesses about how world emissions might change over the coming century, under alternative, plausible, consistent assumptions about trends in population growth, economic growth, and technological change. The scenarios were developed to provide emission inputs for model projections of how the climate is likely to change. But emission scenarios can also be constructed to serve a different purpose than as inputs to climate models. They can instead be derived from alternative environmental goals, in order to examine the goals' feasibility and evaluate alternative ways of achieving them. The most widely proposed goal for limiting climate change is to slow and stop the present increase in atmospheric CO_2 abundance, stabilizing it at some specified higher level. Figure 4.3 shows five emission scenarios constructed to stabilize atmospheric CO_2 at 450, 550, 650, 750, and 1000 p.p.m., respectively. (Recall that

CO_2 has increased from about 270 to 380 p.p.m. over the past 200 years, and is presently increasing by about 2 p.p.m. per year.) These emission scenarios all have a similar appearance, with emissions initially rising, then turning and declining sharply. The lower the CO_2-stabilization target, the sooner this reversal occurs. Stabilizing at 550 p.p.m., for example, requires emissions to peak at 11 GtC around 2035, then decline to 7 GtC by 2100 and 3–4 GtC by 2200. Stabilizing at 1000 p.p.m., on the other hand, allows emissions to grow until nearly 2100, with a slow decline over the following two centuries.

But recall that CO_2 is just one of several greenhouse gases emitted by human activities, albeit the most important one. When considering multiple greenhouse gases, it is convenient to express them in terms of an equivalent amount of CO_2 alone that gives the same climate warming. For example, one analysis found that stabilizing CO_2 at 540 p.p.m. with middle-of-the-road assumptions about growth of other greenhouse gases would produce as much warming as 750 p.p.m. of CO_2 alone. In other words, stabilizing climate change at a level equivalent to 750 p.p.m. of CO_2 would require limiting CO_2 itself to 540 p.p.m.. Similarly, stabilizing total warming at a CO_2-equivalent of 550 p.p.m. would require stabilizing CO_2 itself around 400 p.p.m.

The scenarios in Figure 4.3 are not the only way to reach the specified CO_2 stabilization targets. There are many paths to reach each target, including some that start cutting emissions immediately and others that make larger cuts starting later. But the shape of the scenarios in Figure 4.3, in which early emission growth is followed by sharp later reductions, makes the cost of attaining the concentration targets low. There are four reasons this shape tends to reduce costs: it avoids premature scrapping of long-lived capital equipment such as power stations; it allows more time to develop new low-emitting technologies; it allows time for the natural carbon cycle to help remove early emissions from the atmosphere by the time the concentration target becomes binding; and by delaying emission-reduction expenditures, it reduces their present value through discounting.[2]

Whether emission scenarios are constructed as projections or as goals, they are only a starting point for thinking about mitigation policy. Projection scenarios like those produced for the IPCC show alternative guesses about how emissions might grow, but say little about what might cause actual emissions to follow one of these paths rather than another. Target-based scenarios show what emission paths are consistent with various atmospheric-stabilization goals, but say little about what efforts might be required to put emissions on one of these paths. Answering these questions requires asking what factors cause emissions to change in one way

---

[2] See Appendix A1 for a discussion of discounting.

rather than another, and what measures are available to deflect their trends. We now turn to these questions.

4.2.2   Factors underlying emission trends

A first step to understanding causes of emission trends and ways to reduce them is to decompose emission trends into trends in the underlying factors discussed in Chapter 3: population, economic growth (GDP per person), and technology (CO_2 emitted per dollar of GDP). These factors influence each other, of course, so this mathematical decomposition does not mean that each factor can be varied independent of the others. Over the past few decades, world population growth has declined to slightly more than 1 percent per year due to sharp reductions in fertility rates through much (not all) of the world, while world GDP per person has grown on average between 1 and 2 percent annually. Technological change has tended to reduce emissions, such that CO_2 emitted per dollar of GDP declined slightly more than 1 percent per year on average over the twentieth century. The net effect of these trends is that CO_2 emissions grew by about 1 percent per year on average over the twentieth century.

Any policy to slow or reverse emission growth would have to achieve some combination of shifts in trends of one or more of these underlying factors: a faster decline in world population growth; slower economic growth; and an acceleration of technological innovation. But policies explicitly aimed at limiting population or economic growth are contentious in the extreme. All emission scenarios discussed above assume continued decline in world population growth. But because the causes of recent fertility decreases are only weakly understood, there is real uncertainty over how effective policies to accelerate the trend could be, even disregarding the deep political, cultural, and religious controversies associated with such policies. And of course, if future fertility declines should slow or reverse, emission growth could be well above even the top of the IPCC scenario range.

Explicit policies to limit economic growth for the sake of the environment are, if possible, even more explosive a topic than policies to control population growth. Relieving the extreme poverty of many of the world's citizens provides a compelling rationale for worldwide growth in economic output and incomes, provided the income growth actually reaches those who need it. But even in the world's richest societies, a central focus of public policies remains the continued promotion of economic growth, and there is no clear evidence that people's desire to consume more exhibits any satiation, with the single exception of food. Both within countries, and globally between countries, economic growth also fulfils a sharper political need: the prospect of continued growth can mute political pressure to resolve inequities and social problems, by giving the disadvantaged hope that their lot will improve if they wait. Although it has long been suggested that aggregate material consumption must eventually be expected to cease growing – by writers ranging from classical economists to modern ecological economists and development theorists who propose a long-term slowing of growth in the richest economies and gradual convergence of global inequalities – this argument has secured little traction in policy debates.

As a result, to the extent that policies target any of the components of emission growth, the focus is nearly always on technology. Technological advances that lead to reductions in either how much energy the economy consumes, or how much CO_2 is emitted to produce this energy, could greatly reduce future greenhouse-gas emissions. The huge gap between the A1F and A1T scenarios illustrates the vast divergence of futures potentially attainable through alternative technological paths alone.

4.2.3   Technological options to reduce emissions

There are several broad areas of technological change that can contribute to reducing greenhouse-gas emissions: increasing the efficiency of energy use; developing and improving non-fossil-fuel primary energy sources (renewable sources like solar, wind, and biomass, as well as nuclear fission and fusion); and carbon sequestration, by which CO_2 from fossil-fuel combustion is captured and stored in biological or geological reservoirs rather than released to the atmosphere. While fossil-fuel CO_2 is the biggest area to seek emission-cutting opportunities, significant reductions can also be achieved in CO_2 emissions associated with land-use practices, for example by protecting or re-growing forests, or through agricultural practices to increase carbon accumulation in soils, and in reducing industrial emissions of the other major greenhouse gases – methane, nitrous oxide, and halocarbons.

Historically, the world economy shows a clear long-term trend of decreasing emissions of CO_2 per dollar of economic output (GDP). This decline has reflected both decreased energy use per GDP in industrial economies, and a gradual shift from higher carbon-emitting energy sources (wood, then coal) toward lower-carbon sources (petroleum and natural gas, with some movement toward nuclear power and renewable sources), reducing CO_2 emitted per unit of energy consumed.

Energy use per GDP at the national level has at times decreased as fast as 2 percent per year, but the periods of fastest decline have reflected special conditions – either periods of unusually fast shift in the mix of economic activity away from more intensely energy-consuming sectors (like steel) toward less energy-intense sectors (like services), or responses to energy price shocks. Consequently there are serious doubts whether such high rates can be sustained for decades or extended to the whole world. For the whole world over the twentieth century, the average decrease in energy use per dollar of GDP was more modest, about 1 percent per year. The energy efficiency of most major industrial equipment has been improving for decades, so while further improvements remain likely, they are likely to be gradual. More recently, there have been large efficiency improvements in certain areas of energy use, through expansion of such innovations as high-efficiency fluorescent lighting and low-emissivity windows. Large further gains in end-use energy efficiency are possible, but not necessarily easy, because many areas face socio-economic obstacles in addition to technical ones. For example, many of the largest opportunities for improved automobile fuel economy involve changes in vehicle characteristics such as size and acceleration that consumers have resisted.

How fast can energy use per GDP decline over the twentyfirst century and beyond? Most analysts project that the average twentieth-century rate of about 1 percent per year can be sustained, while some suggest that with stronger incentives from higher energy prices, sustained decreases of about 2 percent per year are possible. Because efficiency improvements are usually realized by installing new capital equipment, however, a high rate of efficiency improvement can be sustained only under conditions of strong economic growth.

The other technological route to reducing emissions is to reduce the amount of CO_2 emitted per unit of energy generated. Some such reduction can be accomplished in the near term by shifting from higher to lower-carbon fossil fuels, i.e. from coal to natural gas. Larger reductions, however, would require shifting to energy sources that emit no CO_2 – renewable sources or nuclear – or to new technologies that allow extracting energy from fossil fuels without emitting the CO_2 to the atmosphere.

Renewable energy sources already provide several percent of world primary energy. The two biggest sources – firewood and hydroelectricity – cannot be expanded much more, however, while the remaining sources – solar, wind, geothermal, ocean thermal, and tidal – together contribute less than 1 percent of present world energy. Wind and solar power are already cost-competitive in some niche applications, principally remote locations, while large-scale modern wind turbines are increasingly competitive even in centralized power systems if sufficiently windy sites are available. Continuing incremental innovation to increase conversion efficiencies and reduce costs can allow further expansion of these sources. There are also some prospects for expansion of commercial biomass energy based on cultivating fast-growing crops on plantations and efficiently burning them. This source can provide energy with no net CO_2 emissions if the plantations are managed sustainably.

All these renewable sources suffer from two problems, however, that are likely to obstruct their expansion to a large fraction of world energy use. First, they have low power densities, so providing a lot of power requires installations covering huge areas. For example, to meet a substantial fraction of world energy needs with solar power would require a solar array of some tens of thousands of square kilometers. Biomass is limited by photosynthesis to even lower power density, about 0.6 W/m^2, so meeting a substantial fraction of world energy needs this way would require about as much land area as is presently used for agriculture worldwide. Second, most renewable sources generate energy only intermittently – when the Sun is shining or the wind is blowing – so they need backup or energy-storage systems to provide reliable delivered energy all day and all year. While substantial expansion of renewable sources is possible, these two issues represent serious challenges to their providing a large fraction of world energy.

The renewable source that would best avoid the problems of low power density and intermittency would be solar energy collected on large arrays in space and transmitted to receiving stations on Earth. Because sunlight is more intense and always available in space, only about one-tenth the array area would be needed than on the Earth's surface to deliver the same power. Despite this advantage, the cost of launching material into orbit is so high that the cost of delivered energy from such a source has thus far been judged prohibitive.

Like renewable sources, nuclear fission and fusion are energy sources that emit no CO_2 to the atmosphere. Nuclear fission reactors, which generate energy by split-ting uranium or plutonium atoms, have been in large-scale use worldwide for decades. Construction of new reactors has been stalled since the 1970s, however, due to concerns over safety, waste disposal, terrorism, and the risks of nuclear-weapons proliferation from diversion of reactor fuel. Fusion reactors, which gen-erate energy by fusing two hydrogen atoms to create a helium atom, remain in development after decades of research.

Nuclear power could make a large contribution to world energy by mid-twentyfirst century, subject to major remaining obstacles and uncertainties. For fission, new reactor designs hold the promise of greatly improving safety, while the waste-disposal problem appears likely to be solvable technically, if perhaps not politically. The most acute challenge to large-scale expansion of fission remains the risk of contributing to proliferation of nuclear weapons, which may not be surmountable. It has also been suggested that world uranium resources may not be adequate to sustain a large-scale fission industry without chemical reprocess-ing of fuel, a process likely to further increase the risk of illicit diversion of fuel to make weapons. Fusion remains a speculative resource, still awaiting the tech-nical breakthroughs that must precede commercial viability, so no significant contribution from it can reasonably be expected for at least several decades.

A final major technical direction to large-scale reductions in CO_2 emissions is to burn fossil fuels, but in a way that releases little or no CO_2 to the atmosphere. The approach, called carbon separation and sequestration, involves decomposing the fossil fuel before it is burned into its major chemical constituents, hydrogen and carbon. The hydrogen is burned to provide energy, producing emissions that consist primarily of harmless water vapor. The carbon is buried in a long-term reservoir underground or undersea. Recent progress in these technologies suggests that this approach is technically viable, is compatible with present energy systems, and would cost substantially less than present renewable or nuclear sources. The crucial question about this approach is the long-term stability of the CO_2-storage sites. If the stored carbon escapes back to the atmosphere too fast – on average, if it returns faster than a few thousand years – then this approach would be ineffective. Early research suggests that some sequestration sites, including depleted oil and gas fields, deep salty aquifers, deep coal seams, and perhaps the deep ocean for certain chemical forms of carbon, are reliably stable for much longer periods. Although the safety and stability of these reservoirs needs further research and careful assessment of associated risks, it appears at present that CO_2 sequestration has substantial promise to reduce emissions, particularly over the next few decades when fossil fuels remain the primary source of world energy.

Carbon can also be sequestered biologically, in trees or soils, although both the potential size and longevity of these reservoirs, as well as their vulnerability to changed environmental conditions (for example, warming can speed the decomposition of dead plant matter and consequently the return of stored carbon to the atmosphere) make them look less promising than geological sequestration. Systems that combine growing biomass for energy production with separating and sequestering the resultant carbon, as well as systems that recapture CO_2 directly from the atmosphere, also show some early promise.

Although progress is occurring in many areas of energy technology, rapid acceleration of progress will be required for any of these sources to make a serious contribution to the climate-change problem. One recent analysis of the amount of carbon-free energy required under various scenarios illustrates the size of the challenge. If energy demand lies near the middle of the IPCC scenario range, then stabilizing CO_2 at 550 p.p.m. would require having more carbon-free energy sources on-line by 2050 than today's total world energy consumption. If the stabilization target is lower, then even more carbon-free energy must be available by that time: stabilizing CO_2 at 450 p.p.m. would require twice as much carbon-free energy by 2050 as total world energy consumption today, while stabilizing at 350 p.p.m. of CO_2 would require 2.5 times as much. These projections depend, of course, on how fast energy demand grows with no mitigation efforts. If it grows slower than the middle of the IPCC range, less carbon-free energy will be needed; faster growth means that even more will be needed.

If the environmental goal is expressed as a limit on global temperature rise rather than CO_2 concentration, then the amount of carbon-free energy needed will also depend on a second key uncertainty, the climate sensitivity. Recall from Chapter 3 that global climate sensitivity – the global-temperature response to a doubling of pre-industrial CO_2 – is estimated to lie between 1.5 and 4.5 °C. If climate change is to be limited to some specified change in global-average temperature, then larger emission cuts – and consequently more rapid development of non-CO_2-emitting energy sources – will be needed if climate sensitivity lies near the top of this range than if it lies near the bottom. For example, one analysis set a limit of 2 °C total global warming, and found that if climate sensitivity was low (1.5 °C), annual emissions could rise to 14 GtC before starting to decline after mid-century, but if sensitivity was high (4.5 °C), stringent cuts must begin immediately and annual emissions must drop below 1 GtC by 2050.

This is a sobering picture. Unless we are extremely lucky – meaning that climate sensitivity lies near the bottom of the accepted range, **and** unconstrained energy demand growth lies near the bottom of the IPCC scenario range, **and** no serious climate surprises or abrupt changes occur – then even meeting limits that allow a great deal of climate change to occur will require unprecedented acceleration of the development of new carbon-free energy technologies.

4.2.4   National policy responses

How can public policy help speed the development and deployment of non-emitting energy sources? The decisions to develop and adopt emission-reducing technologies, like most economic decisions, will mostly not be made by governments for the purpose of slowing climate change. Rather, they will be made by thousands or millions of individuals and organizations for their own diverse purposes, responding to their own perceptions of their present opportunities, costs, and risks, and their guesses about future ones. Government policy can, however, influence these millions of choices by private actors, by enhancing their capacity to make socially preferred choices, providing them with information to facilitate such choices, or changing their perceptions of the opportunities, costs, and risks that motivate their choices. There are four major types of public policy relevant to greenhouse-gas emission mitigation: direct public expenditures, conventional regulation, market-based regulatory instruments such as emission taxes or tradable emission permits, and various initiatives that rely on information, education, and voluntary actions.

The most important role for direct public expenditure in mitigation policy is in government-supported research and development (R&D) of advanced energy technologies. There are strong arguments for public investment in energy R&D – both as a means to facilitate emission reductions, and to correct the market failure that arises from the "public-good" character of R&D, whereby private firms invest too little because they cannot reap the full public benefit of the knowledge gained. The magnitude of the energy-technology challenge suggests a central role for government R&D investment, because the long time horizons and large risks involved make it difficult for private firms to make the required level of expenditures. Despite frequent exhortations for technological innovation as the main route to manage the climate-change issue, spending on energy research has been declining for at least the past decade in most industrialized countries. In the USA, several studies and expert commissions over the past ten years have recommended large increases in federal energy R&D expenditures on improving energy efficiency, renewable energy, nuclear energy, and carbon capture and sequestration, but the recommended increases and shifts of priorities have not taken place (see, for example, National Commission on Energy Policy, 2004).

Conventional environmental regulations specify some performance target that each emitter – typically a firm or a factory – must meet. Targets can be defined in various ways, including maximum emissions of a pollutant per year, or per unit of operations (for example, regulations for automobile exhaust are defined in terms of grams of each pollutant emitted per mile driven), or maximum concentration of a pollutant in emissions. A few environmental regulations have not just required performance targets but also specified particular technologies or processes to achieve them, but this type of policy is much less common. Conventional regulation of greenhouse-gas emissions would involve limiting the emissions of specific plants or types of equipment.

Regulations of this type have been responsible for much of the environmental improvements achieved over the past 30 years. Conventional regulation has been increasingly criticized, however, for imposing higher costs than are necessary to achieve a specified environmental benefit. There are two principal reasons for this inefficiency. First, when performance targets are imposed uniformly across some class of emission sources (for example, each plant must cut its emissions by 20 percent), this may result in large variation between emitters in the marginal cost of control.[3] When marginal costs differ between emitters, it is possible to gain the same environmental benefit at lower cost by shifting control efforts among sources, cutting more where cuts are relatively cheap (i.e. where marginal control costs are lower) and less where they are more difficult and costly (where marginal control costs are higher). Second, such standards give inadequate incentives for emission-reducing technological innovations, because much of the benefit of such innovations comes from lowering the cost of reductions beyond the standards. Although these criticisms have sometimes been overstated, they are substantially correct.

[3] See Appendix A2 for a discussion of marginal costs.

The principal response to these inefficiencies of conventional regulation has been market-based regulatory mechanisms, of which there are two major forms: emission fees, and tradable emission-permit systems, sometimes called "cap-and-trade" systems. Under an emission fee, each source must pay a specified charge for each unit (say a ton) of pollution emitted. Under a cap-and-trade system, the government distributes permits, each of which allows the holder to emit one unit of pollution (again, a ton), and which may be bought and sold among emitters.[4] The advantage of these systems is their flexibility. They do not specify how much any particular source must cut; each source may choose how much to emit, as long as they either pay the emission fee or hold a permit for the amount. In either case, the effect of the policy is to put a price for the emitter onto each ton they emit, which motivates them to reduce their emissions in order to avoid the cost. If the policy is set at the right level – the level of the fee or the number of permits distributed – and if emitters respond optimally to these incentives, then the socially optimal configuration of emissions occurs, both in terms of how much is emitted overall and how much is coming from each individual source.

These two forms of market-based policies, and hybrids of the two, are the most frequently proposed policies to limit greenhouse-gas emissions. A greenhouse-gas emission fee, or "carbon tax," would be applied to fossil fuels in proportion to their carbon content. For ease of administration, the fee would probably be applied at the point of import or primary production (the coal mine or oil well), with a credit generated whenever fuel is diverted into a non-emissive use such as petrochemical manufacture or long-term sequestration. The fee would then stay in the price of the fuel as it passes through the economy, raising the cost of all goods and services that use carbon-based energy. A cap-and-trade system would be implemented in a similar way: a permit for the carbon content would be required to import or to extract a unit of fossil fuel, while a new permit would be generated for each unit of carbon sequestered in a stable reservoir. The cost of the permit, like the emission fee, would follow the fuel through the economy, raising the price of carbon-based goods and services. Alternatively, an emission fee or a permit requirement can be imposed at the point where the fossil fuel is burned and the CO_2 emitted. This is the approach being taken by the EU emission trading scheme. For administrative feasibility, this approach can only be applied to a relatively small number of large, stationary sources of emissions – for example electrical generating stations and large industrial facilities – thereby limiting the economy-wide effect and concentrating it on fewer goods.

There are a few important differences between emission fee and tradable-permit systems. An emission fee must be paid on every unit of pollution that the

---

[4] See Appendix A3 for a discussion of emissions trading.

emitter releases to the atmosphere. Consequently, emission fees transfer wealth from emitters to the government. A permit system would make the same wealth transfer if, as usually proposed, emitters must buy their permits in an auction. In existing permit systems, however, permits are usually not auctioned, but are distributed at no charge to current emitters. Implemented in this way, tradable-permit systems are much less costly to current emitters and consequently meet less opposition and are more frequently adopted. Moreover, the two systems operate quite differently when there is uncertainty about the costs and benefits of cutting emissions. A permit system fixes the total quantity emitted, regardless of how much it costs to reduce to that level. A tax system fixes the cost of the last unit of emission to be cut – because emitters will cut until it is cheaper to pay the tax than cut further, then stop – regardless of how much emissions are actually reduced to reach this point. Consequently, if we are uncertain about the costs or benefits of cutting emissions, whether we prefer a tax or a permit system depends on which of these quantities – the marginal cost, or the total quantity reduced – it is more important to get right.[5]

Hybrids of emission-fee and tradable-permit systems are also possible, which combine some aspects of each system to jointly manage the risks of both regulating too strictly and not strictly enough. One example of a hybrid system would be an emission fee that is not charged on all emissions, but only on emissions above some baseline (and rebated for emissions below the baseline). Alternatively, a tradable-permit system can include a "safety valve," a provision to relax the emission target by issuing more permits if the price at which the permits trade exceeds some pre-established limit. An emission-fee system can also include a similar provision to adjust the fee upward if emissions or their rate of growth exceed some threshold, or downward if emissions fall faster than expected.

In order to avoid near-term disruption, either an emission-fee or a tradable-permit system would have to be phased in gradually, but with a commitment to progressive increases in stringency – raising the fee or decreasing the supply of emission permits – over time. A pre-announced schedule of future increases would allow emitters a stable planning environment and assure them that investments in reducing their emissions will pay off. On the other hand, the system would need some flexibility to respond to new knowledge about climate change or technological developments that called for either stronger or weaker efforts. A basic challenge in designing a practical mitigation policy is how to balance

---

[5] For somewhat technical reasons, this means that a tax system is preferred when abatement costs increase sharply as controls are tightened, and a permit system is preferred when marginal damages of pollution increase sharply as controls are loosened to allow more pollution.

these needs, providing enough flexibility to respond to new information and also enough firmness to assure emitters that investments in mitigation will pay off.

A final category of public policy tools includes educational, information-based, and voluntary measures. These measures seek to influence emissions by educating citizens and emitters about the climate issue and available opportunities to reduce emissions. They may seek to help guide beneficial choices, or to motivate, coordinate, or provide recognition for voluntary mitigation efforts by firms. For example, policies requiring that cars and appliances carry labels explaining their energy efficiency allow consumers to consider this factor in their purchasing decisions. Providing climate forecasts can also motivate people to consider the need to adapt to a changing climate.

These measures can increase the effectiveness of other, stronger forms of mitigation policy by helping private actors recognize, understand, and respond to the incentives embodied in other policies. In addition, policies of this type can sometimes deploy real incentives. For example, requiring firms to report their emissions publicly can motivate managers to make significant efforts to reduce them, as can voluntary programs that carry prizes, public recognition, or linkages to government purchasing. But there are limits to the effectiveness of these policies, particularly when they must stand alone. Voluntary and informational measures usually cannot motivate changes that carry substantial costs or require large investments.

4.2.5   International policy responses

This discussion of policy options has focused at the national level, because this is where direct regulatory authority over emitters lies. But because greenhouse-gas emissions anywhere contribute to climate change everywhere, and because no one country dominates world emissions, mitigation efforts must be coordinated internationally to be effective. Since there is no world government that can enact or implement authoritative policy, international coordination requires negotiation among representatives of national governments, usually with industry and environmental groups and other non-state actors present and trying to influence the outcome.

As discussed in Chapter 1, present international policy for climate change is weak. The Framework Convention on Climate Change (FCCC) states a demanding and appropriate long-term goal, but contains few and weak concrete measures to pursue it: a few procedural and reporting obligations, and a non-binding emission target that most parties failed to meet. The Kyoto Protocol imposes binding near-term national emission limits, and authorizes international flexibility mechanisms (similar to the tradable-permit systems discussed above) that seek to reduce overall cost by allowing voluntary exchanges of emission-reduction obligations between countries. But the Protocol's prospects have been weakened by its late entry into force, the American decision not to ratify, and the fact that not all nations that have ratified have enacted policies strong enough to meet their targets. The Protocol has also been attacked, with some justification, for its emphasis on arbitrary near-term national emission targets with no means of implementing them, rather than the larger, longer-term reductions that will be required to effectively limit climate change.

The form of national commitments

While national policies can impose obligations directly onto emitters, international policies are limited, with few exceptions, to imposing obligations on national governments. Like domestic policies, these national commitments to international policies can take several broad forms. Governments can commit to performance targets, such as limits on national emissions. Alternatively, they can commit to enact national policies, such as emission fees. Or they can submit to international processes such as reporting and exchange of information, assessment of national policies, or reviews of their progress, which seek to motivate and enable national actions. These three approaches are similar to three of the approaches to domestic mitigation policy discussed above, conventional regulation, market-based mechanisms, and voluntary and information-based approaches. Governments also have the additional choice of establishing international institutions to which they can delegate some level of authority to enact or implement policies.

National emission targets are the most widely used form of international mitigation commitment. They have been used in both the FCCC and the Kyoto Protocol, and in several other important environmental treaties. National targets have the advantages of being simple, clear, and familiar. In addition, by defining clear responsibilities but leaving the means of implementation up to national governments, targets allow governments to be held accountable for their commitments with minimal intrusion on their sovereignty.

But national targets also have serious disadvantages. Because the great majority of emissions come not from governments directly but from their citizens and enterprises, a national target has no concrete effect until it is implemented in some form of domestic mitigation policy. But the effects of domestic policies are uncertain, so governments cannot know in advance how costly and difficult it will be to meet a given emissions target, or even whether it is within their power to do so. Even at the domestic level, decades of failure to attain the air-quality standards in the US Clean Air Act in many regions illustrate the potential gulf between adopting a target and achieving it. This uncertainty about targets' attainability introduces a tension in attempts to negotiate effective targets, between making targets demanding and creating strong incentives to achieve them. If the consequences of a nation's missing a target are minor, there is little incentive for serious efforts to meet it. But if the consequences are severe, governments will likely only agree to targets that they are highly confident they can meet – i.e. targets that are weak or that include broad loopholes. Moreover, while clear, demanding targets can act as powerful motivators, their motivational power is concentrated at the boundary between meeting and not meeting them. Incentives to exceed a target you already expect to meet, or to narrow the gap when you are clearly falling short, are much weaker.

Making national targets into an effective policy tool for greenhouse-gas mitigation will require maintaining the motivating power of challenging, attention-getting targets, but also providing incentives for emission reduction over a wide range – both above and below the target – while also making the costs of missing the target small enough that governments are willing to adopt ambitious targets. An approach that can serve these goals is to combine targets with mechanisms that allow flexibility in how the target is achieved. These mechanisms can provide national targets with some of the cost-reducing advantages of market-based mechanisms in domestic policy. Three types of flexibility have been proposed, called "what," "when," and "where" flexibility, all of which are represented to at least some degree in the Kyoto Protocol. "What" flexibility allows nations to choose how to distribute their mitigation effort among the activities and greenhouse gases that contribute to climate change. The Kyoto Protocol does this by defining a single national target for total greenhouse-gas emissions, with each gas counted in proportion to an estimate of its relative contribution to climate change, allowing reductions in any gas to count toward the commitment. "When" flexibility allows nations to allocate their mitigation effort over time to contribute to the long-term goal of cutting emissions. The Protocol includes a little of this form of flexibility, by defining mitigation commitments over the five-year period from 2008 to 2012 rather than year by year, although this still represents a much shorter-term focus than the actual time horizon of the climate problem. "Where" flexibility is the form that corresponds most closely to tradable emission permits in national policy. It allows nations to make voluntary exchanges of their mitigation obligations with each other, so a nation facing high mitigation costs can cut less than its commitment, and instead pay another nation with lower mitigation costs to cut more than its commitment. Calculations of mitigation costs suggest that such flexibility to shift mitigation effort between nations can reduce the overall cost of meeting a specified global target by as much as 80 to 90 percent, while "what" and "when" flexibility offer additional significant cost savings.

When applied at the international level, however, such flexibility mechanisms – particularly "where" flexibility – pose policy design and implementation challenges greater than in domestic policy. The most important problems concern designing the connection between the mitigation obligations held by national governments and by individual emitters within and across national borders, and ensuring the integrity of a trading system that operates across diverse national policy systems. These and related design problems are not yet adequately addressed in the Kyoto Protocol, and are probably hard and complex enough that any such system would have to be operated, evaluated, and adjusted – perhaps repeatedly – throughout the life of the agreement.

A second potential approach to national commitments is to negotiate the stringency of national policies rather than targets for national emissions. For greenhouse-gas mitigation, the major proposal is to negotiate the level of a fee or tax on greenhouse-gas emissions. In principle such a fee could be applied to all greenhouse gases, but most proposals concentrate on the carbon emitted from fossil fuels – an international carbon tax. Negotiating over carbon taxes rather than emission targets avoids the problem of uncertain achievability, since it is clearly within national governments' power to impose taxes. Under a tax, we would expect emitters to cut until the cost of an additional unit of reduction exceeded the tax rate. As discussed above, the level of the tax would consequently determine the marginal cost of mitigation, but the actual quantity of mitigation achieved would be uncertain.

For a tax to achieve the large emission reductions needed to stabilize atmospheric CO_2 concentration, a firm commitment would be needed to raise the tax to high levels over time, even if the rate is initially set low. Unless emissions respond much more readily to the tax than is presently estimated, a carbon tax big enough to achieve the required reductions would represent a large shift in the basis of government finance in every participating country, as well as a large wealth transfer. This would face all manner of domestic opposition, and pose serious problems of implementation and fiscal planning. In addition, the complexity and diversity of national tax systems would require intensive negotiations over implementation to prevent national loopholes, such as exemptions or other special treatment granted to carbon-intensive export industries. An internationally negotiated carbon tax is also more likely to raise sovereignty-based objections than negotiations over emission targets. Hybrid systems combining elements of both emission taxes and tradable-permit systems, such as have been proposed for domestic policy, could help to manage some of the uncertainties associated with either system separately and ease some of the negotiation problems.

A third proposed approach to international mitigation commitments is for national governments to negotiate international procedures, institutions, and rules, rather than committing to either specific policies or specific emission targets. This approach was seriously considered in the climate negotiations of the early 1990s, under the slogan "pledge and review." Under this proposal, governments would pledge to enact mitigation policies and measures of their own choosing, and state what results they expected. They would then be subject to periodic international review of the design and implementation of their policies, the results they projected, and the results they achieved. This approach was proposed to meet the objection that national governments would never accept binding international policy commitments because they would infringe on their sovereignty. Pledge-and-review sought to create and exploit less formal and stringent, but still potentially effective, incentives for national policy-makers, such as wanting to appear competent and responsible, and avoiding international criticism and embarrassment. In negotiation of the Kyoto Protocol, pledge-and-review was subsequently abandoned in favor of binding national emission targets.

Like voluntary and information-based measures in domestic policy, this approach can effectively complement other forms of international commitment. Indeed, such procedural commitments may be necessary to promote effective implementation of other forms of commitment. But as in domestic policy, exclusive reliance on this approach poses significant problems. A key weakness of the approach is that it gives the strongest incentives to those governments and officials who care the most about the issue and their reputation, who would be most inclined to control their emissions in any case, while providing less effective incentives to those who care less. While a procedural approach clearly has some merit and should be intelligently exploited, there are two major questions about its suitability as the main approach to international policy. First, can it deploy strong enough incentives for national policy-makers to motivate the required cuts in greenhouse-gas emissions? Second, are its proponents correct that pursuing stronger commitments from national governments on the climate issue is futile?

A final approach is to create international institutions with real authority to enact mitigation or other climate-related policies directly. Often proposed by those who are most skeptical of the competency or resolve of national governments to address climate change, this approach draws on the historical example of the negotiated formation of real trans-national power in the European Union, and on the specter of some potential future environmental calamity that would demand strong, perhaps even dictatorial, leadership. But the approach has serious problems. Excluding the possibility of coup d'état or revolution, the only way such authority could be established would be by negotiated agreement among national governments. But if these governments cannot muster the will to enact specific policies adequate to manage the climate issue, how can they be expected to agree to give away their authority on the issue completely to some international body? Moreover, it is not clear how such an international authority, once created, could be controlled to ensure its competency and democratic accountability. Effective international authority in specific issue areas can grow over time through successive negotiated decisions. The present international trade regime, the World Trade Organization (WTO), has evolved this way. The WTO provides a plausible model for a climate regime that would combine agreement on targets, policies, procedures, and international institutions to which specific authorities are progressively delegated as experience is gained and nations' confidence in the process grows. But it is not practical to contemplate this approach as a single, large-scale decision that would resolve the present international deadlock on climate.

Implementation and review

Whatever form of commitments national governments adopt, there remains the question of how to make them follow through with their commitments once they have made them. Many environmentalists advocate "treaties with teeth," meaning commitments backed up by punitive sanctions for non-compliance, such as trade restrictions, fines, or withdrawals of aid or investment. But such sanctions are difficult to use effectively. They are hard to negotiate, and hard to agree to use in a specific case even once agreed in principle. They are frequently arbitrary or illegitimate in the way they are deployed, and are frequently ineffective at changing the target country's behavior even when they are deployed. In view of these problems, international environmental agreements usually avoid punitive sanctions, and instead rely on softer methods of persuasion such as reporting and review processes, proceedings that focus more on collegial problem-solving than on identification and punishment of non-compliance, or other means of exerting moral pressure on national policy-makers. For the distinct problem of parties who want to meet their commitments but are unable to do so, these approaches can be augmented by various forms of financial and technical assistance.

How effectively commitments are implemented influences what commitments governments are willing to make, because what each government is willing to do depends on their confidence that others will fulfill their respective commitments. Consequently, procedures to monitor implementation deliver two types of incentives for governments to meet commitments: not just the risk of being embarrassed if they fail to meet their own commitments, but also increased confidence that others are meeting theirs. Effective implementation procedures can consequently increase governments' willingness to agree to commitments, but only if these enjoy strong international support in principle. Strong implementation procedures cannot be used by a few activists to get the rest of the world, especially the major powers, to participate in the first place if they are not interested in doing so. Consequently, while negotiations have begun to develop procedures to review compliance with the climate regime, these are unlikely to contribute much to the problem until a critical mass of major nations have adopted some form of substantive commitments that they clearly intend to take seriously.

Sequencing of commitments

The global scale of the climate issue means that all of the world's major emitting nations must eventually participate in effective mitigation commitments. But an effective global mitigation regime cannot be constructed in one step; rather, it must be phased in over time. There are two elements to the required sequencing. First, the stringency of commitments will have to be revised over time, both to reduce costs and disruptions by phasing in changes gradually, and to adapt over time to new information and changed conditions. Second, participation will also change over time, most likely with an initial core of participants that subsequently broadens toward global participation. The sequencing of both commitments and participation should be guided by similar principles: early actions should make an effective contribution to the problem, should make subsequent expansion of efforts easier, and should not lock in some restrictive or inferior form of policy.

There have been several controversies over the most effective way to sequence participation and stringency of commitments in an international climate regime. The most serious controversy has concerned whether all major nations, industrialized and developing, must undertake simultaneous mitigation commitments from the outset. The provisions of the Framework Convention, and several compelling arguments based in both fairness and practicality, all support initial mitigation commitments by the rich, industrialized countries, with the developing countries beginning mitigation efforts later. But some have argued that this approach, or any approach in which a subset of major world nations undertake deep cuts, will obstruct subsequent expansion of the regime because emission-intensive industries will escape the controls by re-locating in countries not controlling emissions. Such movement of investment would hinder effectiveness of the treaty, because emissions are moved abroad instead of being reduced, and would also obstruct subsequent expansion of the regime by making it more costly for those initially outside subsequently to join. Instead, it is argued that a mitigation regime should first seek to involve most or all nations required for an effective long-term solution to the problem – ever if this means that what these participating nations do is very weak – and only gradually move all nations together toward progressive deepening of requirements.

At its most basic, this argument concerns the relative speed of two dynamic processes: the movement of emission-intensive investment to non-participating nations, and the expansion of participation. The argument presumes that the former process is much faster than the latter. But the limited experience gained so far with building international environmental regimes suggests otherwise. The Montreal Protocol provides the most relevant example. Initially negotiated with stringent CFC controls for a relatively small group of nations – and criticized at the time for the risk that CFC-based production would move to non-participating nations – the ozone treaty moved rapidly both to expand to near-global participation and to increase the stringency of controls. The detailed design of control measures in the treaty, in particular its restrictions on trade in relevant products with non-parties, served both to deter the flow of CFC investment outside the control zone and to motivate additional nations to join. Although the details will be harder for climate, greenhouse-gas mitigation commitments could similarly be designed to minimize incentives for investment to move outside the regime and to promote subsequent expansion.

4.3   Putting it together: balancing benefits and costs of mitigation and adaptation

Considering our knowledge and uncertainty about the impacts of climate change, about potential responses for adaptation and mitigation, and about alternative ways to design policies to promote these, how should we decide what to do? The most widespread approach to evaluating policy decisions is to compare the social benefits of each proposed policy to its social costs. Preferred policies are those that make net social benefits – benefits minus costs – as large as possible.

When a policy can be varied over some quantitative range and the question is where to set it – for example, how much shall we spend on health care or national defense, or how tightly shall we limit emissions of some pollutant – the level that maximizes net benefits is found by looking at marginal costs and benefits. As a pollutant is controlled more tightly, it is usually the case that marginal costs increase and marginal benefits decrease: cutting the first ton gains a large benefit for a low cost, cutting the second ton brings a little less benefit for a little more cost, and so on. Social benefits are maximized by controlling up to the last unit that brings a benefit larger than its cost, i.e. to the point at which the marginal cost and marginal benefit of control are equal.

A response to climate change must balance two kinds of effort, mitigation and adaptation. Consequently, the principle for maximizing social benefits is somewhat more complicated, in that there are three marginal quantities that should all be equalized. Emissions should be cut to the point where the marginal cost of mitigation – the cost of cutting one more unit of emissions – is equal to the marginal damage of climate-change impacts from the last unit emitted. But the marginal damage of these impacts is also reduced by adaptation measures, which should be undertaken to the point where the marginal cost of the last increment of adaptation is also equal to the marginal damage avoided by that increment of adaptation. So the condition that defines a socially optimal policy response is that the marginal costs of mitigation and adaptation, and the marginal damage from the remaining climate change, should all be equal to each other.

This is fine in theory. But using this definition to identify a socially optimal climate-change policy requires quantitative estimates of the costs of mitigation, the costs of adaptation measures, and the costs of the damages of climate-change impacts, over a wide range of possible mitigation and adaptation levels. For all three of these quantities, we have some knowledge, but there is substantial uncertainty.

4.3.1   Estimates of the cost of mitigation

Of the three types of cost, we have the most information about mitigation costs. Dozens of analyses of mitigation costs have been conducted for the USA and other regions, and for the world as a whole. These analyses use economic models that make baseline assumptions about population and economic growth and specify various technical details about the energy sector, such as fossil-fuel resources and technological trends. The models are then used to compare the baseline future to an alternative in which emissions are controlled, either by a limit on national or regional emissions or by a carbon tax. For any specified emission limit, models calculate both the total and marginal cost of achieving that limit, relative to the assumed baseline. The marginal cost connects the analyses of emission limits and taxes; if the emission limit carries a specified marginal cost per ton, then that emission limit could in principle be achieved by a tax per ton equal to that marginal cost.

In these analyses, mitigation costs increase rapidly as emissions are reduced from the baseline. Small reductions, up to about 10 percent cuts from the baseline, come at very small total cost, with marginal costs ranging from zero or even negative values to about $10–$30 per ton of carbon.[6] Costs increase with steeper

---

[6] Translating costs per metric ton of carbon reduced into price increases for consumers involves some approximation, because it depends on how final energy is produced from primary energy in the economy. A reasonable approximation is that a cost of $30 per metric ton of carbon would raise the price of gasoline by 7 or 8 cents per US gallon, gas-generated electricity by 0.3 cents per kWh, and coal-generated electricity by 0.8 cents per kWh.

emission reductions, but differ substantially between studies. For example, model studies of the effects of the Kyoto Protocol to the US economy have found total costs ranging from less than 0.2 percent to more than 2 percent of GDP, with associated marginal cost running from about $15 to more than $200 per ton of carbon. As emissions are cut further, the divergence among estimates grows wider. In controlled model comparisons, the estimated cost of stabilizing atmospheric CO_2 at 550 p.p.m. ranges from about 0.1 to 1.7 percent of world economic output in different models, while stabilizing at 450 p.p.m. costs from 1 to 4 percent of world output (world output that is projected to be 4–9 times higher in 2050 than in 1990). In addition to varying between models, estimated costs of stabilization vary two- or three-fold depending on the emission trajectory leading to stabilization, and vary four-fold depending on the assumed baseline emission scenario (stabilizing at 550 p.p.m. costs four times more starting from the highest emission scenario as a baseline than from the lowest).

This wide range of estimates might seem to suggest we know very little about the cost of mitigation, but this impression is somewhat misleading. Much of the variation in mitigation-cost estimates comes from different assumptions about the nature of the policy imposed. For example, large cost differences come from whether nations are allowed flexibility in meeting their mitigation commitment. Allowing maximum flexibility to shift mitigation among different gases, from place to place, and over time reduces estimated costs more than ten-fold. In addition, a substantial cost difference arises from different ways of returning the money collected by a carbon tax to the economy. Costs are lower if funds are recycled efficiently, particularly by reducing some existing tax on labor or capital. These differences in cost estimates therefore do not reflect uncertainty, but rather provide guidance on how to design low-cost mitigation policies.

Some of the differences between mitigation-cost estimates, however, do reflect real limits to our knowledge about relevant characteristics of the economy. The two most important uncertainties are linked to each other. The first concerns the baseline, the path emissions will follow with no mitigation efforts. The second concerns the case of substitution in the economy, especially through the availability of emission-reducing technological innovations – i.e. how strongly emission-reducing innovation will take place both with and without changes in policies and prices. These uncertainties are linked because the baseline incorporates assumptions about the rate of technological innovation without any explicit policy to reduce emissions, while the cost of any reduction from a given baseline depends on how readily innovation responds to policy-induced incentives. There is suggestive evidence from the history of other environmental issues that technological change is quite responsive to policy, in that estimates of the costs of meeting environmental targets made in advance tend to be significantly higher than the costs that are actually realized. These uncertainties suggest the possibility that the rate of innovation, and its responsiveness to policy, might itself be capable of being influenced by policy. Substantially larger public R&D investment in non-carbon technologies and support and facilitation for technology assessment could help lower barriers to commercialization of emission-reducing innovations. Similarly, policies that give private actors incentives to commercialize emission-reducing innovations by placing some cost on emissions, even a small one, could both advance knowledge about the responsiveness of innovation and make it more responsive.

4.3.2   Estimates of the cost of adaptation and climate-change impacts

In addition to mitigation costs, taking a cost–benefit approach to climate policy also requires estimates of the costs of climate-change impacts and of adaptation measures to reduce them. Few of these are available, and most of them are crude and preliminary. Most impact studies have sought only to describe impacts – which is hard enough – rather than evaluate them in dollar terms, but monetary estimates are needed to allow quantitative balancing of benefits and costs. For impacts on goods or services that are exchanged in competitive markets – for example, changes in production of agricultural products or commercial timber – the market value of projected changes is a reasonable estimate of total social harm or benefit. But many impacts are projected to affect resources and aspects of the environment that clearly matter to people, but for which only limited markets exist, or none. Examples would include the satisfaction New Englanders derive from their climate and the landscape that depends on it, such as bright, snowy winter days, colorful fall foliage, and forests that support maple syrup production. These impacts are difficult to estimate, in part because different people or societies might perceive quality-of-life to depend on different aspects of climate, and in part because people's preferences might shift over time. People might adapt their preferences to climate change as it occurs, learning to live with – or even to like – the environment and climate that they have, even if this is a New England without snow or maple trees, as long as the climate stays stable long enough for such preferences to form.

Some studies estimate the value of such non-market effects by asking people how much they would be willing to pay to protect these assets, but these studies are relatively crude and the methodology contentious. An alternative to expressing all impacts in dollar terms is to keep track of impacts using several metrics. For example, one proposed approach counts five dimensions of impacts: market impacts in dollars, human lives lost, biodiversity loss, income redistribution, and some estimate for quality-of-life changes. Even this five-dimensional accounting scheme may be too simplified to reflect the full range of people's valuation of climate change, however, especially in estimating climate-related quality-of-life.

In view of these difficulties, quantitative estimates of the value of climate-change impacts are fewer, cruder, and less well founded than estimates of mitigation costs. But weak or not, quantitative estimates of the aggregate impacts of different climate futures are needed to assess policy choices in cost–benefit terms. Among the studies that attempt such monetary aggregate impact estimates, most are highly simplified and judgmental, even arbitrary. They typically sort economic sectors according to their presumed degree of climate sensitivity, assign judgmental estimates of projected losses in sensitive sectors, and add them up. Early calculations using this approach estimated upper-bound damages ranging from 1 to 2.5 percent of GDP for a doubled-CO_2 (550 p.p.m.) equilibrium. Other studies then took these point estimates for doubled-CO_2 and sketched cost curves as functions of temperature change, using various simple functional forms. Attempts to assess comprehensive damages from greenhouse-gas emissions in terms of marginal costs per ton of emissions have spanned a wide range, from $1–2 per ton carbon through more than $300. Most estimates lie between $10 and $100, and depend strongly on the discount rate used: studies using discount rates around 5 percent give estimates that fall near the low end of this range, while those using rates around 1 to 2 percent fall near the high end.

Only in a few specific economic sectors have climate-impact studies gone beyond the crude analyses described above. In particular, impacts on agriculture and from sea-level rise have both been comprehensively studied because these sectors have two important advantages for producing monetary estimates of damages. First, good computer models are available to estimate how climate change will affect crop productivity and sea level. Second, markets for agricultural products, farmland, and coastal land provide a reasonable basis for pricing the changes. But even in these areas, projected dollar values of impacts vary strongly from study to study.

4.3.3   Integrated assessment of climate-change adaptation and mitigation

Analyses that consider climate-change impacts, adaptation, and mitigation together are called integrated assessments (IA). Integrated assessment models represent the climate system, the socio-economic factors that drive emissions, the impacts of climate change, and potential mitigation and adaptation responses in a consistent quantitative framework. While highly simplified, these models can be used to simulate the effects of different mitigation and adaptation strategies, and to calculate the costs and benefits of alternative scenarios and policies. In addition, they allow the comparison of marginal benefits and marginal costs that are needed to identify optimal policies, rather than simply calculating the costs of a specified policy.

In view of the complexity involved in combining all these components in an integrated assessment model, each component is typically represented in highly simplified form. This is especially true for the representation of climate-change impacts. Because of the uncertainty of physically grounded impact projections, and the fact that they are available for only a few domains of impacts, integrated assessments have mostly not used them, relying instead on crude judgmental estimates of the monetary value of aggregate climate impacts.

Several of the best known analyses using integrated assessment models have concluded that little near-term mitigation is warranted. One study, for example, found that the optimal level of mitigation over the twentyfirst century was only about a 10 percent reduction from the projected emission baseline. This small reduction could be achieved with a carbon tax of only $5 to $20 per ton, and generated a net social benefit of only about 0.04 percent of world output, relative to no mitigation.

There are several reasons that the optimal level of mitigation in these analyses is so small. The most important of these include how they discount future costs and benefits, their representation of technological change, and their weak treatment of uncertainty. Under the baseline emissions and climate scenarios assumed, large climate-change impacts only begin to appear around 2100 and later. Discounted back to the present using a conventional approach (for example, a constant 5 percent discount rate), the costs of these impacts become insignificant. Mitigation efforts to deflect late-century climate change must be taken early in the century, however, so their costs are much less reduced by discounting. Moreover, most of these analyses assume that technological change will proceed rapidly, but is quite unresponsive to policies or prices. Consequently, while mitigation and adaptation grow substantially cheaper as time passes, this progress does not require price or policy incentives and cannot much be accelerated by them. Since it requires early mitigation to avoid impacts many decades later, these technology assumptions strongly favor adaptation over mitigation, just as discounting does.

These assumptions are not unreasonable, but nor are they clearly correct. And changing any of them can sharply increase the desired level of mitigation. Changing the discounting of future costs and benefits, by using either a lower discount rate or non-exponential forms of discounting, favors substantially more twentyfirst-century mitigation. Increasing the assumed responsiveness of technological innovation to price and policy incentives also favors more mitigation, as does introducing uncertainty into projections of climate-change impacts. If a significant probability of abrupt or catastrophic climate change is assumed, greater mitigation is favored if we are risk-averse. We are frequently risk-averse in policy, defending against risks with relatively low probability if their consequences are bad enough, for example in defense and public health policy.

Mitigation can be further preferred if some decisions made under uncertainty are irreversible. In any analysis of decision under uncertainty, irreversible actions carry a penalty: to take an irreversible action, you require a larger expected benefit than if the action was reversible. Since each ton of CO_2 emitted makes a change in the climate that is irreversible for centuries or longer, we wish to emit less than a simple comparison of our best guesses of costs and benefits would tell us – thus favoring more mitigation. Putting emissions into the atmosphere is not the only irreversibility that might be present, however. Major capital investments in non-emitting energy technologies are also only slowly reversible, since you cannot simply change your mind and recover invested funds with no penalty. This latter irreversibility tilts the decision the other way, favoring less mitigation than we would make based on best-guess costs and benefits. But to the extent that the distribution of potential climate-change impacts includes even a small possibility of severe or catastrophic changes, the balance of irreversibilities tends to favor more, not less, mitigation effort.

In sum, integrated-assessment models provide a valuable framework for thinking through the large-scale structure of climate response, in particular for balancing mitigation and adaptation efforts over time. But at present their results are merely illustrative. They are highly sensitive to alternative specification of uncertain aspects of the evolution of the economy – for example, regarding technological change, productivity trends, ease of substitution in the economy, and the nature of decision-makers' responses to changes in policies and prices – as well as to alternative specifications of the design of policies and alternative treatments of discounting. The results showing that optimal mitigation is small depend on a collection of assumptions that strongly favor this result. Alternative plausible specifications can increase the calculated optimal reduction in twentyfirst-century emissions from less than 10 percent to more than 80 percent. With such a wide range of plausible alternatives in optimal policy, it is clear that while these models provide a valuable structure for examining policies and identifying key uncertainties for research, they as yet provide little practical guidance for the choice of a specific mitigation policy.

4.4   A third class of response: geoengineering

In addition to mitigation and adaptation, a third class of potential responses to the climate issue involves actively manipulating the climate to offset the effects of increased greenhouse gases in the atmosphere. This approach, usually called *geoengineering*, includes a diverse collection of proposals. Some involve obstructing incoming sunlight, by injecting reflective aerosols into the stratosphere or launching screens into space to block a small fraction of the Sun's light from reaching the Earth. Other proposals involve manipulating the global carbon cycle, for example to increase CO_2 uptake by the ocean by fertilizing marine plankton with some limiting nutrient such as iron, or by directly removing CO_2 from the atmosphere. Several geoengineering approaches appear promising in early studies, although any such active manipulation of the Earth on planetary scale inevitably carries large, perhaps unanticipated, risks.

The greatest strength of most geoengineering approaches is that they can modify the climate much faster than mitigation can. While even an extreme program of mitigation would take several decades to begin altering the climate, a Sun shield in space (which might take about a decade to develop and launch) would begin offsetting the effect of elevated atmospheric CO_2 as soon as it was deployed. Consequently, geoengineering approaches could provide some protection against a worst-case scenario of rapid, severe climate change. The scale of effort required could likely be achieved by one or a few rich and technically advanced nations. Such projects would pose serious legal, diplomatic, and political problems, such as conflict over who has the authority to undertake them, whether they would impinge on existing international treaties or require new ones, and how the burden should be shared. Most problematic would be the potential for conflict between nations proposing such a project and those opposing it, either because of principled opposition to active planetary-scale manipulation or because the opponents expect the climatic effects of the project to harm them. In the extreme, a geoengineering project could be regarded as a hostile act by another state, akin to Cold War proposals to use active weather modification as a weapon.

There is no indication at present that advantageous mitigation and adaptation possibilities have been exhausted, so there is no basis for expecting that actual deployment of geoengineering approaches will, or should, play any significant near-term role in our climate response. Still, and despite its large evident challenges, this class of approaches merits serious consideration, so options can be available and their risks better characterized in case they are needed.

4.5   Conclusion: policy choices under uncertainty

The most basic question in choosing a response to climate change is how to act responsibly under uncertainty. Although uncertainties in climate change are not as overwhelming and debilitating as some advocates suggest, they still pervade every part of the problem. Future paths of CO_2 emissions have wide uncertainty. So do estimates of the quantitative magnitude of the global climate response and the resultant regional-scale climate changes and impacts. So too do estimates of the efficacy, costs, and other effects of the various approaches to mitigation and adaptation.

About some of these matters we know quite a lot, about others we know less, and about some we know very little. Some of these uncertainties derive from limits in our knowledge of Earth systems. Much uncertainty also derives from our limited ability to predict human behavior and patterns of development. While much progress in knowledge can be expected, neither or these types of uncertainty will be reduced to insignificance anytime soon.

Consequently, widespread uncertainty will remain a central element of the climate-change problem, which will not fade into insignifance with another decade (or two or three decades) of delaying concrete response while doing further research. As in any high-stakes domain of human affairs, choices must be made despite continuing uncertainty, and responsible choices require balancing prudent near-term actions, continuing efforts to learn more, and provisions to adjust and refine our choices as conditions change and as we do learn more. As a general matter, these principles of decision-making under uncertainty are widely accepted, but how to apply them to the climate-change issue has been contentious in the extreme. In the next and final chapter, we will present our views on a responsible and practical path forward for responding to the climate-change issue, based the present state of knowledge and the prospects for advancing it.

Further reading for Chapter 4

Hoffert, M. I. *et al.* (2002). Advanced technology paths to global climate stability: energy for a greenhouse planet. *Science*, **295**, 981–987.

> An analysis of the present status of the major types of technology that could reduce or replace present carbon-emitting energy sources, with some illustrative – but very sobering – calculations of how rapidly these non-emitting energy sources must expand within the next few decades to achieve atmospheric CO_2 stabilization at roughly double the pre-industrial level.

IPCC (2001b). *Climate Change 2001: Impacts, Adaptation, and Vulnerability. Contribution of Working Group II to the Third Assessment Report of the Intergovernmental Panel on Climate Change*, ed. J. J. McCarthy, O. F. Canziani, N. A. Leary, D. J. Dokken, and K. S. White. Cambridge and New York: Cambridge University Press.

> This is the most recent full assessment of the IPCC's Working Group II, which summarizes present knowledge about potential impacts of climate change, ability to adapt, and vulnerability of environmental and social systems to climate change.

IPCC (2000c). *Climate Change 2001: Mitigation. Contribution of Working Group III to the Third Assessment Report of the Intergovernmental Panel on Climate Change*, ed. B. Metz, O. Davidson, J. R. Swart and J. Pan. Cambridge and New York: Cambridge University Press.

> This report provides the IPCC's most recent comprehensive review of technical options and policies for reducing greenhouse-gas emissions. It includes detailed discussion of costs, of barriers to expansion of mitigation, and of analytic methods and models for assessing mitigation options.

IPCC (2000). *Emission Scenarios: Special report of the Intergovernmental Panel on Climate Change*, ed. N. Nakicenovic and R. Swart. Cambridge and New York: Cambridge University Press.

> This special report provides the results and background for the IPCC's most recent set of emission scenarios, which were used as inputs to the climate-model projections in the Working Group 1 report and are summarized in this chapter.

Pacala, S. and Socolow, R. (2004). Stabilization wedges: solving the climate problem for the next 50 years with current technologies. *Science*, **305**, 968–972.

> An alternative view, more optimistic than that of Hoffert *et al.*, of the capacity of presently available technologies to achieve the emission reductions needed over the next 50 years to move toward stabilization of atmospheric CO_2 concentration.

Parson, E. A. and Fisher-Vanden, K. (1997). Integrated assessment models of global climate change. *Annual Review of Energy and the Environment*, **22**, 589–628.

> A review of the major issues involved in building and using integrated assessment models of global climate change, with summaries and evaluations of the major contributions to the field.

National Commission on Energy Policy (2004). *Ending the Energy Stalemate: A Bipartisan Strategy to Meet America's Energy Challenges.* December, 2004. Available at www.energycommission. org

> This is the most recent and most comprehensive of the senior panel reports that have examined US energy policy, presenting detailed background studies and recommendations pertaining to climate change, energy supply and security, infrastructure, and energy technology R&D.

US Global Change Research Program, National Assessment Synthesis Team (2001). *Climate Change Impacts on the United States: The Potential Consequences of Climate Variability and Change.* New York: Cambridge University Press.

> This report assesses potential climate-change impacts, vulnerabilities, and capacity for adaptation for the United States. Separate studies examine effects of climate change on nine major US regions and five sectors of national importance. Like the IPCC reports, this assessment involved the work of hundreds of scientists and was subjected to a rigorous and thoroughly documented process of peer review. It is often referred to as "The US National Assessment".

Wigley, T. M. L., Richels, R. and Edmonds, J. A. (1997). Economic and environmental choices in the stabilization of atmospheric CO_2 concentrations. *Nature*, **379**, 240–243.

> The analysis that identified scenarios in which emissions initially rise and then fall sharply as low-cost ways to attain various alternative targets for stabilizing atmospheric concentrations of CO_2.

5

The present impasse and steps forward

In the previous chapters, we have summarized the present state of knowledge and uncertainty on climate change. We have reviewed what is known about the climate and how it is changing, the evidence for a human contribution to the observed changes, and the range of potential changes projected for the coming century, as well as the weaker state of knowledge about potential climate-change impacts and responses. This final chapter is more political, in two senses. First, we present a detailed examination of the present deadlocked politics of the issue, reviewing both who is lining up where, and what arguments are being advanced that are contributing to the current deadlock. Second, we present our own judgments of what kind of response to the climate issue appears to be appropriate, prudent, and practical in view of present scientific knowledge and political alignments.

5.1 Climate-change politics: present positions

Although climate change has been on policy agendas for more than a decade, progress on the issue is stalled both domestically in the United States of America and internationally. As discussed in Chapter 4, there are several broad types of response to the climate issue, including mitigation (reducing emissions), adaptation, and geoengineering. But present controversy, and the present deadlock in policy-making, are nearly exclusively concerned with mitigation – whether to take near-term policy action to reduce emissions, and if so, how stringently and of what form. Mitigation is the principal focus of controversy because it is the form of response for which near-term decisions are most clearly required, and the form of response that most centrally involves the exercise of governmental regulatory authority over private actions. Both adaptation and geoengineering, correctly or not, are widely perceived as being decisions to be made in the future, or as less intrusive into citizens' lives and choices.

In the USA, there has been little progress since President Bush's 2001 decision not to ratify the Kyoto Protocol. As discussed in Chapter 1, current US policy emphasizes research and voluntary initiatives, with the weak aim of reducing national greenhouse-gas intensity – emissions per dollar of GDP – by 18 percent between 2002 and 2012. Given projected growth of the economy, this aim corresponds to a 12 percent *increase* in emissions over that period. Legislation proposed in 2003 (the "McCain–Lieberman bill") included somewhat stronger measures, capping aggregate emissions from the largest sources (about 85 percent of total US emissions) at their 2000 levels in 2010 using an emission trading system. In an October 2003 vote in the United States Senate, this bill was defeated 55 to 43, a surprisingly strong showing in view of the Senate's prior hostility to mitigation measures. In the absence of effective federal action, more than half the States have begun developing mitigation strategies, several of which include comprehensive reduction targets for statewide emissions. California has proposed the measures of greatest potential importance, using its unique authority to regulate motor vehicle pollution to require phased reductions in vehicle CO_2 emissions beginning in the 2009 model year, reaching 30 percent cuts by 2014 (although this regulation is currently facing legal challenges).

At the international level, the USA withdrawal wounded the Kyoto Protocol, but it remains unclear whether the wound is mortal or not. Protocol signatories made substantial progress over the few years following its signature, culminating in the agreements reached in 2002 in Marrakech, in defining the rules for the first round of mitigation commitments in 2008–2012. These agreements were followed by the ratifications of most major industrial nations, then in November 2004 – after several years of ambiguous and contradictory statements – ratification by Russia, which put the Protocol over the required emissions threshold and allowed the protocol to enter into force in February 2005. Most industrialized nations have adopted policies to pursue their Kyoto commitments, although the strength of these measures and their likelihood of success vary greatly. In the strongest positions are the UK and Germany, which are both likely to reduce emissions beyond their Kyoto commitments by substantial margins due to a combination of fortunate circumstances and strong policies. The EU as a whole is making progress toward its promised 8 percent reduction, albeit more slowly. Its aggregate 2002 emissions were 2.9 percent below the 1990 baseline. In addition, the EU has adopted both national targets for member states and – reversing a long-standing skeptical stance toward flexibility measures – an emission-trading system for large sources that will begin in 2005. Japan's policy to pursue its required 6 percent reduction, announced in April 2005, relies on sectoral reduction targets (especially for industry and electrical generation), backed up by a combination of voluntary measures, standards for building efficiency and renewable energy, and government purchasing of low-emissions vehicles, although there is also substantial reliance on sinks and purchased credits. Of the major industrial-country parties, Canada is the least likely to meet its targeted cut of 6 percent, having ratified late following a decade of strong emission growth and unproductive domestic consultations on mitigation measures. Even optimistic official projections suggest that mitigation measures identified and implemented to date are only likely to achieve about one-third of the reductions required under the Protocol.

On the crucial questions of extending and strengthening mitigation commitments beyond 2012, including expanding their scope to the developing nations, essentially no progress has been made. The industrialized and developing nations remain sharply divided on the linked questions of what mitigation commitments developing nations will take on, how soon, and how much financial and technological assistance they will receive in return. Several major oil-exporting states remain adamantly opposed to any mitigation and demand that they be compensated if any is enacted, while the small island and low-lying coastal states, several of them threatened by inundation from sea level rise, continue to plead for aggressive mitigation and assistance, to little effect. Overall, there is no serious international deliberation underway over how to develop an effective mitigation regime over the crucial window of the next few decades.

Among non-government actors, environmental groups predictably favor early mitigation efforts and most support the Kyoto Protocol, including US ratification. Equally predictably, most industry groups oppose the Protocol, although industry positions on broader questions of mitigation and climate policy are more mixed and more interesting. The most active opponents of any near-term mitigation are the major fossil-fuel producers, their industry associations, and their affiliated non-governmental organizations – with the conspicuous exceptions of two oil majors, BP and Royal Dutch/Shell, which are instead positioning themselves as environmentally responsible energy companies. Even the other oil majors have shown recent signs of hedging their bets and softening their anti-mitigation rhetoric. Perhaps this reflects a guess that mitigation is inevitable sooner or later and it is best to appear constructive. Alternatively, it may reflect a recognition that all the low-cost, conventionally exploitable oil and gas will be burned in any case, and consequently that the effects of mitigation policy on them will depend on who captures the benefits of the resultant price increases. Coal producers are most threatened by mitigation, except to the extent that rapid expansion of carbon separation and sequestration technologies mutes this threat, and are most forceful in their opposition.

The stance of the major non-energy industrial firms shows somewhat more flexibility toward mitigation. While none has explicitly supported mitigation policies, a few dozen support organizations that recognize the seriousness of climate change and promote constructive responses. These organizations do not support or oppose specific mitigation proposals, but instead stress the importance of any policies meeting certain principles: for example, that they be science-based, cost-effective, international, and applied broadly and equitably rather than singling out particular industries. A few business sectors have moved further toward supporting mitigation – including minor ones like the skiing industry (the "small island states" of the private sector), which endorsed the 2003 McCain–Lieberman bill, and major ones like the insurance and finance industries, which view climate change as both a risk to asset values and a potential market opportunity for new financial instruments. Thus far, however, these industry groups still carry less clout than those opposing mitigation or sitting on the fence. Overall, while an increasing number of leading firms in the USA and more in other industrialized nations are at least muting their opposition to mitigation, these movements remain small and tentative. Without leadership from governments, there are too many business risks for even the most responsible and far-sighted firms to pursue mitigation efforts on their own.

5.2   Climate-change politics: the arguments against action

A glance at the present policy debate shows that many people, including many leading figures in the present US administration, oppose near-term mitigation efforts. Opponents of mitigation advance several arguments to support their position, of three main types: attacks on the specific terms of the Kyoto Protocol; rejections of the scientific evidence for climate change; and claims that large emission reductions will be excessively, perhaps ruinously, costly. In this chapter, we summarize and critique these arguments. We address the specific arguments against the Kyoto Protocol in this section, the arguments against the scientific evidence for climate change in Section 5.3, and the arguments that mitigation is too costly in Section 5.4, where we also present our own outline of a proposed response to climate change.

Because the Kyoto Protocol is the most important current initiative in international climate policy, much of the present debate focuses on the Protocol itself rather than the broader question of what type of response to climate change, including how much mitigation effort, is appropriate. Since the initial negotiation and signing of the Protocol, opponents have argued that the Protocol is "fatally flawed" and must be abandoned, based on several lines of attack. Some of these – for example, attacks on the Protocol's flexibility mechanisms and implementation provisions for vagueness or loopholes – have declined in prominence and persuasiveness in the past few years as these elements of the Protocol have been improved. The principal continuing attacks are directed against the Protocol's emissions commitments themselves, on two grounds: that they do not require actions by the major developing countries; and that they are arbitrary and not based on science. We consider these in turn.

It is correct that the Kyoto targets do not include any developing-country obligations for the first commitment period, nor any guarantee that these countries will accept mitigation commitments in the future. The treaty was drafted this way for both practical and principled reasons. As a practical matter, negotiators recognized that industrialized countries had more technical and financial resources to make initial policy changes and investments, and were willing to make these initial commitments when developing countries were not. It was also widely argued that because past fossil-fuel use by the industrialized countries is responsible for most of the present buildup of atmospheric greenhouse gases, these nations have the obligation to take the lead in slowing the changes. Even before the Protocol negotiations, this widely endorsed normative view had found formal expression in the Framework Convention's Principle of Common but Differentiated Responsibility.

But critics of the Protocol dispute both the practical and normative basis for this approach. They argue that not controlling developing-country emissions makes the treaty ineffective, because emissions-intensive industries will simply move there rather than reducing emissions. This movement of investment will offset any reductions being made in industrialized countries, and will obstruct subsequent attempts to extend mitigation efforts to developing countries. They also argue that excluding developing countries is unfair, because these nations are already both major competitors in many sectors of world trade, and major emitters of greenhouse gases – and will soon grow to make up the majority of world emissions – so letting them not make cuts will put the nations that are cutting emissions at a major competitive disadvantage.

There is some merit in these arguments on both sides. A mitigation regime must eventually achieve near-global participation if it is to be effective, and therefore must not erect obstacles in the path of its own expansion. But this condition does not require full global participation immediately. Rather, a regime that is designed to absorb new members and that provides strong enough incentives for them to join can be viable even if it begins only with a relatively small "coalition of the willing."

Any judgment about how the burden of responding to climate change should be shared will reflect some balancing of past responsibility, projected future responsibility, and capability. Greater historical responsibility and greater financial and technological capacity both suggest that the rich industrialized countries should take the lead in initial mitigation efforts. But the projected surge in capital investment in developing countries, even in the near term, represents a major opportunity to limit future emission growth by shifting this investment toward lower-emitting technologies. Moreover, as the economic growth projected over the next few decades in the developing countries increases both their incomes and their emissions, their responsibilities to contribute to global mitigation will also increase.

With this large disparity between past responsibility and projected future responsibility, reaching agreement in advance on how to share future contributions to global mitigation may well be impossible. Rather, the appropriate balance of effort can probably only be resolved in the evolution of the regime over time, not in a once-for-all bargain or formula. The Framework Convention's principle seems to capture the required approach about as specifically as could reasonably be negotiated in advance: all must participate, but how much depends on their degree of capability, responsibility, and concern. Because these will all shift over time, the distribution of contributions to mitigation will also have to shift over time. Expanding participation over time to include mitigation efforts in the developing countries will be essential, although this does not necessarily mean that developing countries will pay the entire cost of these: there are various ways that the location of mitigation can be separated from who pays for it. But demanding global participation from the outset virtually assures an extended deadlock in which no action is taken. Indeed, sustaining the present deadlock might well be one of the unstated goals of this debating tactic.

The second major charge against the Protocol's emission limits is that they are arbitrary, not based on science, and paradoxically, that they are both too strong and too weak: too strong and thus too costly in the near term, but too weak to achieve any serious reduction of global climate change. Like the criticisms of the Protocol for excluding developing countries, the factual basis for these changes is largely correct, but their implications for action are not clear. In particular, they are not sufficient to conclude, as the critics do, that the Protocol must be rejected. Rather, reaching this judgement requires considering what the likely alternative course of action would be if the Protocol were abandoned, and whether this would be preferable.

The Protocol's targets are arbitrary, because they reflect a bargained compromise between some nations that sought stricter targets and others that sought weaker ones or none. In this respect they resemble all politically negotiated outcomes: arbitrariness is no special weakness of the Kyoto Protocol. Nor are the targets "based on science," because scientific knowledge cannot specify any particular level of emission target. Scientific knowledge can help inform decisions about targets, by projecting the consequences of alternative emission levels – faster climate change from weaker emission controls, slower climate change from stronger ones. Science might even help to identify emission paths associated with a large increase in the risk of abrupt climate changes – although this would require substantial advances from present knowledge. But without some confidently known environmental threshold that everyone would agree must be avoided, no emission target is more or less "based on science" than any other. This charge could be leveled equally against any emission target, and so is essentially meaningless.

It is also largely correct that the Kyoto Protocol's mitigation targets are too strict in the near term and too weak in the long term. Perhaps they were not too strict at the time they were negotiated in 1997, when they allowed 11–15 years to meet the targets: this is debated. But in the eight years since then, only a few nations have made serious efforts to meet the targets. With less than three years left to the start of the commitment period, nations whose efforts are not well underway face a choice between three highly unattractive options: adopting a high-cost, crash program to cut emissions rapidly (if this is even possible); violating their commitments; or relying predominantly on purchased emission credits, thereby weakening the global reductions that are actually achieved. At the same time, the Protocol lacks longer-term targets or measures after 2012, or even specific principles or guidelines for negotiating these, making it ineffective at managing climate change over the required multi-decade time horizon.

These are serious flaws, which must be corrected if the Protocol is to become an effective international instrument for managing climate change. Can they be corrected? The Protocol's flaws are widely recognized by its supporters as well as its opponents, and the most severe ones are likely to be improved – as some already have been – through further negotiations. This is how effective environmental treaties work: they do not resolve environmental issues once and for all, but progressively refine and improve their management over time. If the Protocol is retained, it must evolve as other treaties do, adapting and changing as old problems are resolved and new ones are identified, as political and economic conditions change, and as scientific knowledge and technological capabilities advance.

Many attacks on the Protocol have ignored this evolutionary character, however. They pretend that the present treaty will persist unchanged, to strengthen the case that it should simply be abandoned. In fact, many opponents of the Protocol oppose *any* near-term mitigation effort, but focus their attacks on the Protocol because its clear flaws – particularly under the ridiculous assumption that it will never be amended – make it a vulnerable target. By implying that the Protocol in its present form is the only mitigation option available, they can use its many weaknesses to appear to discredit any mitigation program.

Noting the illegitimacy of this debating tactic does not, however, answer the question of what to do with the Protocol. Supporters of a serious mitigation effort must still consider whether the Protocol should be kept and improved, or abandoned – a question that turns on comparing the Protocol to the likely alternative if it is abandoned. Is this completed but flawed treaty more likely to help or hinder the required longer-term mitigation efforts – including the expansion of mitigation to include developing countries – considering the risk that starting over could produce an extended deadlock or an even more deeply flawed instrument? We will discuss our views on this question of practical politics in the concluding section.

5.3   The present policy debate: use of scientific knowledge and uncertainty

In contrast to the state of knowledge we summarized in Chapter 3, claims are frequently made in the climate-change policy debate that present scientific knowledge does not provide evidence of serious risks from climate change, and certainly does not justify bearing any significant cost to reduce emissions. These arguments take two forms. First, some advocates dispute nearly every specific point of scientific knowledge that we have summarized. We discuss this tactic in Section 5.3.1. Second, instead of making specific scientific arguments, some advocates argue more generally that climate science is too uncertain, so we should wait for more knowledge to reduce uncertainty before taking costly actions that might turn out to be unnecessary. We discuss this broader argument in Section 5.3.2. In both cases, we explain why in our judgment these arguments should be rejected.

5.3.1   Major claims of the "climate skeptics"

In Chapter 3, we summarized present evidence that the Earth's climate is warming rapidly, that greenhouse-gas emissions from human activities are the predominant cause of the most recent rapid warming, that climate will continue to warm over the next century, and that while the precise rate and regional details of future change are uncertain, the range of projected changes includes some that would carry severe impacts. Each of these points, however, is denied by many policy actors who oppose action on climate change and by a small group of self-styled "scientific skeptics" who provide support for these policy views. These people state that the Earth is not warming; that if it is warming, human activities are not responsible; that future warming, if it occurs at all, will be predominantly due to natural causes and much smaller than present projections; and that climate change is on balance likely to be a good thing for people, for various reasons including people's general preference for warm climates and the effects of elevated atmospheric CO_2 on plant growth. Some advocates even claim that these contrarian views are in fact backed by a strong scientific consensus.

These skeptical arguments are rarely if ever advanced in scientific arenas, but in editorial pages, on the internet, or in policy arenas where more lenient standards for evidence and argument apply. Those arguments can be persuasive, both because they sound plausible to those unfamiliar with the relevant scientific literature and because they are often presented in combination with broader political arguments or effective rhetorical devices, such as painting your opponents as extremist, corrupt, or foolish. The media's tendency to uncritically balance opposing views enhances the effectiveness of these arguments, by seeming to give marginal or transparently false claims the same stature as well supported consensus scientific views. In this section, we discuss a few of the most prominent of these skeptics' claims and explain why they are wrong.

Skeptics' Claim 1. **The Earth is not warming**[1]

Evidence of recent warming rests solely on the surface thermometer record. Such data are contradicted by satellite measurements, which are far more reliable. Satellite measurements show a very small warming trend since measurements began in 1979 – about 0.06 °C per decade, much too small to be noticeable.

Despite the seeming plausibility of this argument, each of its points is either outright false or highly contestable. As we summarized in Chapter 3, the first claim – that the surface thermometer record provides the only evidence of global warming – is simply false. The surface record is the single most important and comprehensive source of evidence for warming, but many other independent data sources – for example glacier retreat, shrinkage and thinning of sea ice, warming of seawater, and many forms of paleoclimatic proxy data – all support a consistent picture of a warming global climate.

The next two points – that the satellite temperature record is more reliable than the surface record, and that it contradicts the warming observed in the surface record – are also both simply wrong. Section 3.1.7 discussed the many uncertainties in the satellite data. Making different assumptions for handling these uncertainties produces widely divergent trends from exactly the same satellite data. There are also uncertainties in the surface thermometer record, of course – also

---

[1] The examples that follow are all composites of arguments advanced in newspaper editorial and op-ed pages and other opinion and policy sources over the past several years. For example, this composite claim is drawn from S. Fred Singer, Letters to the Editor: Bad Data Make Global Warming a Cold Case, *Wall Street Journal* (Eastern edition), Nov. 10, 2003, p. A17, and Cal Thomas, Don't succumb to warming hysteria, *Baltimore Sun* editorial page, June 12, 2002, p. 15A.

discussed in Chapter 3 – but there is no basis for claiming that these are larger than the uncertainties in the satellite trends. In 2000, a National Academy of Sciences committee conducted an in-depth study of the two records, and concluded that the satellite record is not more reliable than the surface thermometer record.

But do the satellite data contradict the warming seen in the surface record, and thereby cast doubt on the reality of global warming? Simple atmospheric physics dictates that warming in the lower atmosphere, which the satellite measures, should be slightly larger than warming at the surface. The earliest analysis of the satellite record, published in the early 1990s by the group at the University of Alabama at Huntsville (UAH), showed no warming in the lower atmosphere, contradicting the rapid warming seen in the surface data (Spencer and Christy, 1990). As the satellite record grew longer and several improvements were made in the trend calculation, the UAH group's calculated trend turned to a warming, although the trend has remained smaller than the observed surface warming. If one considers only the UAH calculation, then a discrepancy between the observed surface and satellite trends does exist. The causes and implications of such a discrepancy, however, are unclear. The 2000 National Academy of Science report examined the discrepancy as it then stood and concluded that the disparity between the surface and UAH-calculated satellite trends "in no way invalidates the conclusion that the surface temperature has been rising."

Moreover, several other scientific groups have recently published calculations of trends using exactly the same satellite data, but using different assumptions in their calculation, and have obtained trends as high as 0.26 °C per decade (see the discussion in Section 3.1.7). With satellite-derived trend estimates now ranging from 0.06–0.26 °C per decade[2] (versus a surface trend of 0.1–0.2 °C per decade), there is no longer any basis for claiming that the satellite data contradict the surface warming, let alone claiming that they invalidate the surface warming trend.

Finally, the point that warming of 0.06 °C per decade is insignificant and can simply be ignored, is a non-scientific judgment, and is highly arguable. In the first place, 0.06 °C per decade is the smallest of several estimates of the satellite trend, and there is no basis for claiming it is more likely correct than the other, higher estimates.[3] But even if the recent trend is this small, this rate of warming is certainly not too small to detect. Whether it is negligible or not depends on

---

[2] This does not include a possible correction for stratospheric cooling that would further increase the trends by 0.05 °C per decade.

[3] It is often claimed that the UAH calculation has been verified through comparison to weather-balloon measurements. However, weather balloons do not provide the gold-standard comparison suggested by this argument. In particular, balloons trends are susceptible to the same kinds of instrument and sampling inhomogeneities that affect the surface and MSU records. See the discussion in Mears et al. (2003) and the exchange in letters by Christy and Spencer, and Santer et al., in Science, 301, 1046–1049 (2003).

whether future warming is likely to be this small, and on the impacts of this warming. In particular, it is unlikely that future warming will be this slow: virtually all climate models suggest that twentyfirst-century warming is likely to be substantially faster than twentieth-century warming.

> *Skeptics' Claim 2.* **The Earth may be warming, but human activities are not responsible**
>
> Even if the Earth is warming, this is probably not caused by human activities. It could be a natural climate fluctuation, perhaps part of the continuing recovery from the "little ice age," the cool period from the fifteenth to eighteenth centuries. Or it could be due to increased intensity of sunlight.

In this argument, the skeptics offer two alternative explanations for the observed warming of the twentieth century. The first is that the warming is a recovery from a global cold-period several hundred years ago known as the "little ice age." This suggestion is weakened by the fact that the little ice age, like the medieval warm period before it, appears to have been predominantly a regional variation in the climate around Europe, rather than a global phenomenon. But even if these variations had been global, this argument assumes that the Earth's climate system has a "normal" state that it pushes back to after unusually warm or cold periods, like a stretched spring returning to its normal length. While this might appear commonsensical, it has no foundation in either the record of how climate has varied or the fundamental physics of the atmosphere. The Earth's climate has no "normal" state to which the climate seeks to return, so there is no reason to expect that an unusually cool period will be followed naturally by a return to warmer conditions. This argument is really a different version of the claim that recent warming is due to internal variability of the climate system. As we discussed in Section 3.2, neither proxy records of past climate nor computer climate models suggest that climate can vary far enough or fast enough on its own (i.e. without human interference) to produce the rapid recent warming.

A related argument often heard in the public debate is that the warmth of the last few decades is not exceptional when compared to the climate of the medieval period – thereby implying that recent warming has a natural cause. The primary evidence supporting the unique warmth of the present period over the last thousand years is the proxy record plotted in Figure 3.8. This record is often referred to as the "hockey stick" plot, because it shows a long gradual decline followed by a sharp upturn in the past century, resembling the blade and handle of a hockey stick.

Several criticisms of this result have been made over the past few years. The first of these, and the most widely circulated in policy circles, was a 2003 paper (Soon and Baliunas, 2003) that reviewed the previously published climate-proxy data and concluded that the twentieth century was probably not the warmest period of the past millennium. In other words, this paper contained no new research but only reviewed previous work, yet it purported to overturn the conclusions of that previous work. This in itself provides reasonable grounds to view the paper with suspicion – especially if the previous work has been repeatedly tested by the scientific community and is generally believed to be correct. In this case, such suspicion is warranted (see the discussion of the paper in Monastersky, 2003). Indeed, the editors of the journal where this paper appeared have since suggested that it has such serious methodological errors that it should never have been published. For example, the paper erroneously treated evidence of past periods that were unusually wet or dry as if they indicated periods of unusual warmth, thereby greatly exaggerating the evidence for past warm periods. Moreover, the paper compared these questionable indicators of past climate not to the extreme warmth or rate of warming of the late twentieth century, but to the average temperature over the entire twentieth century. Since it is only the last few decades of the twentieth century that have surpassed estimated boundaries of natural variability, any comparison of the whole century to earlier periods misses the point: the relevant comparison is with the last few decades of the century. Despite these obvious problems, opponents of mitigation have repeatedly cited this paper as showing conclusively that there is nothing anomalous about recent warming. Even ignoring all the indications of serious errors in the paper, it does not support this conclusion.

More recently, there have appeared some more serious and better-founded criticisms of the mathematical methods used to generate Figure 3.8 from the scores of individual proxy records covering the planet. These recent criticisms (von Storch *et al.*, 2004; McIntyre and McKitrick, 2005) have pointed out genuine questions about the analysis, but their implications for our understanding of climate change are probably minor, for several reasons. First, they have not disproved or even attempted to disprove the primary conclusion of Figure 3.8, that today's warmth is remarkable. Rather, they argue that the uncertainty in Figure 3.8 may be greater than previously estimated, so it is *possible* that it underestimates the warmth of the medieval period (see also Moberg *et al.*, 2005). However, the possibility of errors in the hockey-stick plot has long been recognized, by the paleoclimate research community, the IPCC, and even the scientists who produced Figure 3.8. For example, the 2001 IPCC report (2001a) described the conclusion that the 1990s were the warmest decade in the past thousand years as only "likely," which indicates, in the carefully nuanced language they employed to denote degrees of confidence, as much as a one-in-three chance that the conclusion is wrong. It is unclear that these new criticisms substantially alter this confidence estimate. In addition, the hockey stick is not the only evidence that today's warming is probably mostly caused by human activities. Another strong piece of evidence is that climate models cannot reproduce recent warming unless observed increases in CO_2 and other greenhouse gases are included.

This nuanced explanation is not, of course, what climate skeptics argue. Rather, they state that recent criticisms have destroyed the hockey-stick plot – and, since the entire scientific case for global warming is built on the hockey-stick plot, that these criticisms show that global warming is a scientific fraud. This argument completely misrepresents the true state of knowledge about past climate variability and the origin of recent warming.

The second part of the overall skeptics' argument addresses solar output. Could changes in the Sun's radiation output account for the recent warming? Again as we discussed in Section 3.2, there is evidence that increased solar output contributed to warming in the early twentieth century. But satellite measurements of the Sun's output available since the late 1970s do not show enough variations in solar output to account for any significant fraction of the global climate warming of 0.1–0.2 °C per decade that has occurred over that period.

So these proposed natural causes cannot explain the rapid observed warming of the second half of the twentieth century. On the other hand, recent increases in atmospheric greenhouse gases do provide an explanation for recent warming that is theoretically well founded, and that matches the magnitude and the timing of recent warming well. As we discussed in Section 3.2, climate models that exclude recent increases in greenhouse gases cannot simulate the observed recent global climate changes, while models that include greenhouse gases reproduce recent global trends quite well. With this strong evidence in favor of greenhouse gases as the cause, and no evidence supporting alternative explanations, the strong scientific consensus is now that increases in greenhouse gases are responsible for most of the rapid warming of the past few decades.

Skeptics' Claim 3. **Future climate warming will almost certainly be very small**

Even if human activities are causing the recent warming, temperature increases over the twentyfirst century and beyond will likely lie near the bottom of the projected range, or even below it.

Skeptics make two arguments that future warming will be small. The first is that the sensitivity of the global climate is much lower than presently believed, so the Earth will not warm much even if CO_2 emissions continue to grow.[4] The second

---

[4] Recall that the Earth's climate sensitivity measures the eventual (equilibrium) global-average warming that would follow a doubling of the pre-industrial CO_2 concentration. This is a crucial quantity for the climate-change debate, because if sensitivity is low, any specified increase in atmospheric CO_2 will cause less climate change than if sensitivity is high.

is that present projections of future emissions are too high, and that the only
plausible future trend is that emissions will grow slowly or even decline – even
with no effort to curb their growth.

The accepted range of sensitivity estimates has stayed roughly constant for the
past few decades: 1.5–4.5 °C for a doubling of CO_2 concentration. Most of this wide
range comes from uncertainty about how atmospheric water vapor and clouds
will change as the climate warms. Water vapor is itself a strong greenhouse gas,
and much of the warming predicted by climate models comes from increases in
humidity that accompany the warming from increasing CO_2 – an effect known
as the "water-vapor feedback." Consequently, one way to argue that climate sensitivity is low is to claim that humidity will not increase (or not much) as climate
warms. Water vapor is so important that atmospheric scientists have spent great
effort studying how it is regulated and how it will respond to climate change.[5] The
conclusions of the great majority of this research have confirmed what common
sense suggests: surface evaporation will increase in a warmer atmosphere, leading to increases in humidity. This conclusion remains somewhat uncertain, however, because the comprehensive, global measurements of water vapor that would
definitively settle the issue are not yet available. Consequently, arguments that
the water vapor feedback is very small or even negative are frequently advanced
to oppose mitigation.

At present, the most prominent argument in favor of a small water-vapor feedback is the "iris" hypothesis (Lindzen *et al.*, 2001), which proposes that in a warmer
climate, an increased fraction of the water vapor carried upward in tropical thun-
derstorms will fall out as rain. If precipitation increases more than surface evaporation does, the net result will be that a warmer climate is less humid and less cloudy.
The main evidence supporting the hypothesis is the observation that upper-level
cloud cover over a limited region of the western Pacific Ocean tends to decrease
when the surface is warmer. By assuming that less cloud cover means less humidity, and that this correlation generalizes worldwide, the authors conclude that
the atmosphere will dry out as the surface warms. If this is correct, then climate
sensitivity and projected future warming would lie at or below the bottom end of
the present accepted range.[6]

The iris hypothesis is often cited in policy debates as if it is conclusively established and so demonstrates that present warming projections are much too high.
This argument, however, greatly misrepresents the extent to which the hypothesis is scientifically accepted. Since being published, the hypothesis has been subjected to many additional tests, and has not fared well. No subsequent study

---

[5] See IPCC (2001a), Section 7.2.1.

[6] The iris hypothesis is also a negative cloud feedback: as the surface warms and cloud cover is
reduced, enhanced infrared emission to space puts a brake on warming.

has supported the hypothesis, and several have raised doubts about whether the observed correlation between cloud cover and surface temperature is statistically significant (Harrison, 2002), whether it can be attributed to a climate feedback process (Hartmann and Michelsen, 2002), and whether the hypothesized negative feedback would make more than a small reduction in climate sensitivity even if it were correct (Lin *et al.*, 2004). The proponents of the hypothesis have responded to these criticisms (Bell *et al.*, 2002; Lindzen *et al.*, 2002; Chou *et al.*, 2002). While the issue is not decisively resolved, the iris hypothesis presently has virtually no support in the relevant scientific community. It may gain additional support in the future and come to be accepted in some form, but it is so far from that point now that the present uncertainty range of 1.5–4.5 °C probably does a good job of incorporating any downward influence that the hypothesis will turn out to have. Any claim by policy advocates that the iris hypothesis is a well established result that overturns present understanding of climate sensitivity is an insupportable misrepresentation of present scientific opinion.

Others argue that climate will not change much because present projections of future emissions are too high, and that emissions will in fact grow little if at all. The implication is that explicit measures to limit future emissions are not needed, since emissions will not grow much in any case. This claim is highly optimistic in light of historical experience and present estimates of population, economic, and technological trends. This type of argument has also been a widely used stalling tactic in other major environmental issues: virtually every form of pollution that has been proposed for regulatory controls has been claimed by opponents of controls to be incapable of much growth. Climate change is unlike past environmental issues, however, since even if this claim were true, it would not avoid the need for mitigation efforts. Achieving long-term atmospheric stabilization will not just require stopping emission *growth*, but large reductions from present emission levels.

As with other skeptics' claims, it would be great if this were true: it would be most fortunate if emissions stopped growing without active intervention, just as it would be fortunate if the sensitivity of the climate and environment to human disruptions was actually very small. But the claim that emissions will not grow is advanced as pure assertion, with virtually no rational foundation. Any particular projection of global emissions, low or high, is likely to embed assumptions that appear unreasonable when particular small regions or countries are examined, but these usually have a very small effect on the global total. Worldwide, emissions have grown inexorably with growth of world populations and economies, and are likely to grow even faster if world energy systems shift further toward coal and synthetic fuels as cheap conventional oil and gas decline. The only support for the claim that emissions will not grow much is that over the 1990s, emissions grew substantially more slowly than estimated by the IPCC scenarios. But this discrepancy is for just one decade and says very little about future trends, particularly in view of the large shift toward coal in newly planned power plants over the past few years. The range of emission projections is wide because both very high and very low scenarios are plausible based on present knowledge – and since it is well established that people tend to estimate uncertain quantities too confidently, there is no doubt some chance that emissions will lie even outside this wide range, either above or below. The wide range of possible emission futures gives wide latitude for partisan projections: advocates of stringent cuts can claim the highest projections are the most likely, while opponents of controls can claim the lowest are most likely. In fact, either the top or bottom might turn out to be correct, but present knowledge provides no basis for betting confidently on either of them – and roughly in the middle is probably a better bet than either the top or bottom. Rather, any responsible approach to the climate issue must consider the possibility that any point in the range may turn out to be correct.

It is not possible to address all the erroneous and misleading claims advanced in the climate-policy debate. They are too numerous, and they are also a moving target. The advocates advancing these arguments typically retreat step-by-step, as their current claims shift from being merely unsupported to patently ridiculous. For example, while political commentators and editorials still occasionally claim that the Earth is not warming, most prominent scientific skeptics have retreated from this claim over the past few years – much later than the accumulation of evidence warranted, to be sure, but still an indication of their need to maintain some semblance of scientific credibility. We have also focused on the strongest skeptical arguments, giving less emphasis to those that are not just misleading, but patently false. For example, some skeptics have claimed that the surface thermometer record shows no warming over the continental United States – a brazenly false claim, which requires picking and choosing observations to exclude those that show the strongest warming.

Finally, we should note that there are ample opportunities to use biased, misleading, and erroneous scientific arguments on all sides of policy debates. In preparing this book, we have looked hard for prominent, purportedly scientific claims from environmental activists that are as biased or misleading as those discussed here from skeptics, but there is little to be found. A few individual activists make insupportably strong claims about severe human-health impacts from climate change, including asserting that climate change already occurring is implicated in the recent resurgence of infectious diseases. A few others implausibly exaggerate the technological options presently known that would allow mitigation at zero or negative cost – although technological progress could well turn these present wild exaggerations into future realities. But the climate-change statements of the major environmental groups are quite careful, and there is nothing on the environmentalist side resembling the cottage industry of climate skeptics and supporting organizations publishing their claims. Even the two most prominent recent books by climate skeptics that criticize environmentalists' exaggeration of climatic threats find little to attack. One author could find only a few general statements that climate change is one of the most important environmental challenges (or the most) that society faces – pretty tame stuff, which makes no claim to represent specific scientific knowledge, and which may well be true.[7] The authors of the second book charge unnamed environmentalists with spreading a vision of a "hellish climatic catastrophe," but can find no stronger support for this claim than a few statements by President Clinton and Vice-President Gore that recent extreme weather events *might* be linked to global warming.[8] It is certainly possible to exaggerate environmental risks relative to scientific knowledge, and environmental advocates have sometimes done it. But in the present climate-change debate, the weight of misrepresentation appears to lie strongly with the so-called skeptics and the policy actors who use their arguments to oppose greenhouse-gas mitigation.

5.3.2   Defending the boundary between scientific and policy debates: scientific assessment and policy skepticism

Chapter 2 discussed why the use of unsupported and biased scientific arguments is so widespread in policy debates, even when more legitimate non-scientific arguments could be advanced to support the same policy positions. The reason is that the tactic works: because scientific claims get special deference and respect in policy debates, they are frequently effective at persuading people, particularly when the arguments support the listener's prior policy views or are being advanced by someone whose political values they share. Moreover, the risk of being discredited for advancing weak or false scientific arguments is small, due to the lower standards of evidence and argument in policy than in scientific arenas.

As a result of these incentives, the picture of climate-science knowledge offered by the policy debate has remained contentious and uncertain, even as the actual state of scientific knowledge has grown stronger and more consensual. The supposed scientific arguments being waged in the policy debate do not mirror any present debate among climate scientists, but simply obscure or misrepresent

---

[7] All quoted in Lomborg (2001), p. 258.

[8] Quotes reproduced in Michaels and Balling (2000), pp. 7–9. A subsequent book of the same character (Michaels, 2004) mainly criticizes selected examples of journalists highlighting or exaggerating alarming environmental news, and finds only one borderline example of a misleading statement from an environmental group – a rather shrill 2002 press release from Greenpeace.

settled questions. At the same time, this diversion of the policy debate to specious scientific arguments has stifled discussion of economic and political questions, both positive and normative, on which a vigorous public debate should be taking place but is not.

Paradoxically, the increased prominence of distorted scientific claims on climate change over the past several years may partly be a consequence of the growing strength of scientific evidence for the reality and seriousness of climate change. Like negative political campaigns, misrepresenting scientific knowledge in policy debates is a last-ditch strategy, high in risk but potentially effective, available to a side that is losing. Ten or fifteen years ago, opponents of action on climate change could draw on moderately credible scientific claims that are no longer available to them. As scientific consensus has strengthened on key positive points that most citizens and policy-makers would judge to warrant a serious policy response, those who wish to use scientific claims to oppose action are forced to resort to increasingly tendentious, shrill, or misleading claims, or outright false ones.

A similar process has occurred on prior environmental issues as scientific knowledge converged and the issues shifted from matters of scientific dispute to policy action. Once again, the ozone layer provides the closest parallel to the climate-change issue. In the early 1990s, as a policy consensus developed to eliminate CFCs and several other chemicals, based on a strongly converging – although not complete or perfect – scientific consensus about their contribution to ozone depletion, a fierce backlash appeared that prominently circulated several real remaining scientific uncertainties and anomalies, together with all manner of long-refuted and ridiculous claims, in an attempt to roll back policies. Participants in this earlier ozone backlash included several of the same individuals who have now re-appeared as climate skeptics.

What can be done to limit the scope for partisan distortion of scientific knowledge in policy debates? One approach that is not likely to be effective is exhorting the purveyors of false and misleading scientific claims to be more honest. The powerful incentives to use scientific arguments in policy debates – good ones if you have them, bad ones if you don't – are likely to overwhelm any such attempt at moral suasion. Moreover, even if public exposure destroys the credibility of one or a few egregious liars – which seldom happens – the rewards of this role provide ample incentive for others to step up and take their place.

Rather, two approaches are likely to be more effective in reducing the influence of partisan distortion of scientific knowledge in policy debates – and consequently in reducing the incentives to practice such distortion: encouraging participants in policy debates to be more skeptical in evaluating these claims; and ensuring that authoritative scientific advice is available to policy debates through processes that are credible, legitimate, prominent, and have some protections against partisan attack.

Promoting more skeptical treatment of scientific claims advanced in policy debates is the first step, both for claims that are promoted as "skeptical" and for others, even if this cannot equal the rigor with which new claims are scrutinized in scientific settings. Skepticism toward partisan argument is indeed a virtue and it is ironic that those now advancing distorted scientific claims on climate change call themselves "skeptics," since their success depends on policy-makers and citizens not questioning their claims too closely.

An effective skeptical stance depends on asking questions about the foundation of the claim being made. If, for example, someone asserts that the Earth's climate either is, or is not, warming, it is first necessary to ask whether the source of the claim is both expert and impartial – noting that merely holding scientific credentials does not guarantee impartiality. Is the claim based on a peer-reviewed publication? Has it been verified by additional peer-reviewed studies and widely accepted by the relevant scientific community? Are there opposing scientific views? Who holds these opposing views – how many people, of what level of relevant expertise – and what are the grounds for saying that one view is right and the others wrong? Parties to a policy debate should ask these questions, just as scientists would ask them as part of their evaluation of a scientific claim.

Moreover, in policy debates, it is especially important to be skeptical of your friends. Anyone is most at risk of being misled by deceptive and erroneous scientific arguments that are consistent with their own prior beliefs. If you are generally suspicious of unregulated markets and free trade, mistrust the integrity of corporate management, and support government regulation, you are most at risk of being misled by unsupported claims that an environmental risk is well established, immediate, and grave. If you hold the opposite political views – i.e. you believe unregulated markets and free trade advance welfare, regard corporate management as basically trustworthy, and oppose regulation – you are most at risk of being deceived by unsupported claims that support the opposite conclusions – that an environmental risk is undemonstrated, remote, and probably minor. But the true state of the world, and the true state of scientific knowledge about it, takes no account of political values, yours or anyone else's. Making informed and prudent decisions on environmental issues depends on getting access to this knowledge without a political filter.

But no matter how well policy actors follow this advice, scientific claims cannot be evaluated as carefully in policy debates as they are in scientific settings. Since the central problem of scientific advice to policy is that policy actors cannot independently evaluate contending scientific claims, they must to some extent rely on trust. But how can you decide what individuals, institutions, or processes to trust? There is no foolproof guide, but there are hints. Just as publication and subsequent verification in the peer-reviewed literature provide evidence of credibility, publication in certain other outlets provides grounds for suspicion. Claims that are advanced exclusively or primarily in self-published media (for example the internet or publications of advocacy organizations) or in newspapers, particularly in editorials or other opinion pieces, should be viewed with skepticism. So should any claim that a single peer-reviewed scientific paper represents settled knowledge or, even worse, single-handedly overturns an established scientific consensus. Skepticism is particularly warranted for sources that use polemical language or make personal attacks, that state no limits to the certainty or scope of their claims, or that cannot tell what evidence does, or could, weaken their claim.

The most trustworthy source of scientific information for policy debates, and a more practical source than the peer-reviewed literature itself, comes from official scientific assessment processes. As Section 2.5 discussed, scientific assessment processes synthesize, evaluate, and summarize scientific knowledge to inform a decision or a policy debate, often at the request of relevant governmental or international decision-making bodies. Establishing and maintaining effective assessment processes is the most effective way – along with cultivating policy actors' skepticism about scientific claims – to reduce the scope for partisan distortion of scientific knowledge in policy settings.

The principal scientific assessment body for climate change is the IPCC, whose history we presented briefly in Chapter 1 and whose conclusions we have drawn on throughout this book. In gradual steps, IPCC Working Group 1 has made a series of careful statements that have marked out the advance of the scientific consensus about climate change over the past decade. Their 1995 Summary for Policymakers stated that "the balance of evidence suggests that there is a discernible human influence on global climate."[9] Their 2001 summary strengthened this, to say "There is new and stronger evidence that most of the warming observed over the last 50 years is attributable to human activities," and moreover that the warming projected for the twentyfirst century is "very likely to be without precedent during at least the last 10,000 years."[10]

The organization and operations of the IPCC are similar to the highly successful scientific assessment panel previously established for stratospheric ozone, although with the important difference that governments maintain official control over the IPCC. Although this odd hybrid status of the IPCC – partly a scientific body, but partly under governmental control – initially generated confusion and conflict, the IPCC has subsequently developed procedures that have successfully clarified and managed the boundary between its scientific and governmental

---

[9] See IPCC (1996), p. 4.   [10] See IPCC (2001a), pp. 10, 13.

aspects. Under these procedures, governmental control has little or no effect on the detailed work of the assessment, where expert scientific writing teams have full control over the actual report and its technical summary. Governmental control matters most in formal plenary sessions, where national representatives negotiate the Summary for Policymakers – the shortest and most widely circulated product of each assessment report – line by line.

Because of the number, breadth, and stature of the participating scientists, the criticality and thoroughness with which they review the scientific literature, and the rigor with which their reports are peer reviewed, the IPCC assessments have achieved extremely high credibility and significant influence in the policy debate. There have been essentially no substantive criticisms leveled against the content of the reports. They are widely used as references by scientists working in the field or moving into it, and accepted as authoritative by virtually all policy actors engaged in the issue.

The exception to this near-universal acceptance of the consensus stated in the IPCC is in policy debate in the USA. Here, opponents of mitigation have attacked the IPCC reports and their conclusions. They have, in a sense, been compelled to make these attacks, since not doing so would amount to conceding the reality and seriousness of climate change and thereby giving up the supposed scientific basis for their policy positions. Even in the USA, however, advocates who care about their scientific credibility have rarely attacked the IPCC's substantive conclusions. Rather, they argue that the IPCC's process, and consequently the nuances of language in which its conclusions are expressed, are biased toward an alarmist view of climate science and an activist policy stance.

Such an argument might at first seem reasonable. Given the high stakes of the climate issue and the powerful status of scientific claims in policy debates, many policy actors would wish to exercise political influence over IPCC assessments if they could. But the charge does not stand up to scrutiny. IPCC reports are written by hundreds of scientists from dozens of countries, and reviewed by hundreds more individual scientists as well as member governments. All review comments and authors' responses to them are available for public scrutiny. Given the massive and diverse participation and the transparency of the process, any attempt to bias the report toward someone's preferred conclusion would be both offset by opposing pressures, and severely limited by the open character of the deliberations.

Still, on entering office in 2001, the Bush Administration was sufficiently concerned about charges of alarmist bias in the IPCC that it took the unusual step of asking the National Academy of Sciences to conduct an additional review of the 2001 Assessment. This review reaffirmed the soundness of the IPCC report and its major conclusions, as did a subsequent series of official statements by the

American Geophysical Union and the American Meteorological Society.[11] The 2001 IPCC assessment has probably been subjected to more review and scrutiny than any scientific report in history, and all reviews have supported its conclusions. If any bias operates on the IPCC process, it is scientists' general conservatism in evaluating new claims, which grants a massive, grave authority to the assessments' major conclusions.

A subtler charge of bias against the IPCC has been that while the underlying reports are impartial scientific statements, the Summary for Policymakers – a short, non-technical summary drafted by national representatives in plenary session – misrepresents the full report by exaggerating risks and understating uncertainties and qualifications. There have certainly been a few well-known past occasions when the summaries of other scientific assessments of environmental issues have misrepresented the main assessment report – although these occasions have more frequently involved understating environmental risks than overstating them – so this charge merits a serious examination. But like the broader charge of bias in the IPCC, it has not held up. The National Academy of Sciences review was specifically asked to address this charge, and found that the Summary for Policymakers of the Working Group I report appropriately represented the full report, given the need to summarize a thousand-page document into nineteen pages and simplify it for a non-scientific audience. Indeed, the conservatism that pervades the whole IPCC process is also present in the plenary sessions that compose the summary for policy-makers, despite their more political character, both because the scientific lead authors participate in this stage and because many of the government representatives who participate at this stage are in fact government-employed scientists who also worked on the full assessment.

In sum, for all the difficulties they face, the atmospheric-science assessments of the IPCC are on balance highly credible, and highly effective. Their deliberations have maintained an impressive level of independence from political interference, despite an organizational structure that could readily have threatened such independence. To the extent that true synthesis statements of the state of scientific knowledge about climate change exist anywhere, it is in the IPCC assessments. They – and other scientific assessments that achieve similar quality of participation, deliberation, and peer review – are the "gold standard" of trustworthiness of policy-relevant scientific statements, and policy actors can do no better than to rely on them. The continuance of the IPCC's independence and effectiveness

---

[11] Available at http://www.agu.org/sci_soc/policy/climate_change_position.html, and http://www.ametsoc.org/POLICY/climatechangeresearch_2003.html, respectively.

cannot be taken for granted, however, and policy-makers who want continued access to scientific advice of this quality must be vigilant in defending it.

This discussion applies principally to the atmospheric-science assessments produced by IPCC Working Group I. The same arguments about the solidity of the consensus, and the coherent, prominent authoritative statements of key positive points of that consensus, apply much less to the areas of climate-change impacts and options for adaptation and mitigation covered by IPCC Working Groups II and III. These areas have harder questions to answer, less developed bodies of data and evidence, a much wider range of disciplinary diversity to integrate, and longer causal chains to analyze (for example, socio-economic trends make emissions make global climate change make regional climate change make diverse impacts on ecosystems, resources, and human societies). Consequently, it is more difficult to attain an authoritative consensus declaration about the state of relevant scientific knowledge in these domains than it is for atmospheric science. Moreover, Working Groups II and III also address areas in which it is much more difficult to achieve separation between positive questions, which in principle are amenable to scientific investigations, and normative questions, which are not. Predictably in view of these difficulties, the effectiveness of the reports by IPCC Working Groups II and III has been less than that of Working Group I.

5.3.3   Uncertainty and "sound science"

The specific claims denying the emerging consensus and attacks on the scientific assessment process of the IPCC are both relatively crude tactics. The evidence for and consensus supporting each of the positive points we have summarized, and the credibility of the IPCC, are both evident to anyone who takes a moment to look. More sophisticated opponents of mitigation advance subtler arguments. Rather than disputing any specific points, they argue more generally that the science of climate change is highly uncertain, so incurring potentially large costs to protect against climate change is imprudent and wasteful.

A political strategy memo prepared by a consultant to advise Republican candidates how to address the climate-change issue in the 2004 US elections and subsequently leaked to the press provides a strikingly direct statement of this strategy and its objectives.

> [W]hile the economic argument might receive the most applause at Chamber of Commerce meetings, it is the least effective approach among the people you most want to reach – average Americans ... The typical economic approach taken by most Republicans to oppose many environmental rules and regulations simply does not move Democrats and has only limited appeal among independents ...

The scientific debate remains open. Voters believe that there is no consensus about global warming within the scientific community. Should the public come to believe that the scientific issues are settled, their views about global warming will change accordingly. Therefore, *you need to continue to make the lack of scientific certainty a primary issue in the debate* . . . **The scientific debate is closing [against us] but not yet closed. There is still a window of opportunity to challenge the science** [emphasis in original].[12]

This general argument might be summarized as follows.

The response to climate change must be based on sound science, not on speculation or theory. We must not rush to judgment before all the facts are in. There is too much uncertainty and too much that we do not know about climate change. It would be irresponsible to undertake measures to reduce emissions, which could carry high economic costs, until we know that these are warranted.

Parts of this argument are just rhetorical flourishes, such as the statement that policy should be based on "sound science." Setting these aside, the foundation of the arguments – that there is much uncertainty in present scientific knowledge of climate change – is uncontroversial. As President Bush stated when he announced that the United States would not ratify the Kyoto Protocol, "(W)e do not know how much effect natural fluctuations in climate may have had on warming. We do not know how much our climate could or will change in the future. We do not know how fast change will occur, or even how some of our actions could impact it . . . And finally, no one can say with any certainty what constitutes a dangerous level of warming and therefore what level must be avoided."[13] But the suggestion that uncertainty is overwhelming is highly misleading. As we have shown above, there are many points of climate science on which knowledge is quite advanced, and on several key points – such as whether the climate is warming, whether human activities are primarily responsible, and whether the warming is likely to continue – there is essentially no remaining uncertainty of any significance.

Moreover, the central point of this argument – that certainty about climate change is required to justify taking costly mitigation actions, or alternatively, that some higher level of confidence is required than is provided by present scientific knowledge – is not a scientific argument at all, but a normative judgment about when it is appropriate to make costly efforts to forestall an uncertain risk.

---

[12] The original memo, by Frank Luntz of the Luntz Research Companies, is posted online by the Environmental Working Group, at http://www.ewg.org and discussed in J. Lee, A call for softer, greener language, *New York Times*, March 2, 2003, p. 1.

[13] Remarks by President George W. Bush, White House Briefing, White House Rose Garden, June 11, 2001.

In effect, the argument states that the *status quo*, no mitigation policy, should be retained until it can be demonstrated that mitigation is a superior policy. Moreover, by citing "scientific uncertainty" as the reason for not taking action, the argument implies that the required standard of demonstration is total or near-total elimination of scientific uncertainty.

Those advancing this argument are making an analogy, sometimes explicitly and sometimes implicitly, to two other domains of social decision-making where we require a very high standard of evidence to justify certain decisions: criminal law, and scientific research. The rules of criminal trials specify that the defendant is presumed innocent unless the prosecution succeeds in demonstrating guilt "beyond a reasonable doubt." In scientific research, as we discussed in Chapter 2, when a new hypothesis or result claims to contradict present accepted knowledge, it is not accepted until demonstrated to a highly persuasive standard and repeatedly, stringently verified by multiple, independent scientific groups.

In both these cases, the decision to require such a high standard of demonstration is based on a normative judgment about the relative severity of the two possible kinds of error. In any decision taken under uncertainty, there is always some chance of making the wrong choice. A criminal verdict can err by convicting an innocent defendant, or by acquitting a guilty one; scientific judgment can err by accepting a new claim that turns out to be incorrect, or refusing to accept one that turns out to be correct. Criminal trials demand demonstration of guilt beyond a reasonable doubt, thereby biasing the decision in favor of the defendant, because society has long judged that it is much worse to convict an innocent defendant than to acquit a guilty one. In science, the requirement that new claims be strongly verified reflects a similar judgment of the relative severity of the two possible types of error. Accepting an incorrect novel claim is quite costly, since it can confuse and misdirect subsequent research, and cast doubt on the accumulated body of related prior knowledge. But failing to accept a correct novel claim is less costly, because such rejections are always provisional. A correct claim that is not initially accepted will likely keep accumulating supporting evidence until it meets the standard for acceptance, so the cost of imposing this high standard is simply a delay in accepting the claim until more data are obtained.

The crucial point is that, in both these domains, the decision rules are based on normative judgments about which type of error is worse. The worse we judge a particular type of error to be, the more we try to make it unlikely by biasing the decision-making process against it. In doing so, we willingly accept a heightened risk of making the other type of error, because we judge it to be less bad.

But we use different biases in other areas of social decision-making, reflecting different judgments of how bad it is to err in each direction. In civil law – private suits by one party against another, in which usually only monetary damages or requirements to change behavior are at stake – there is no clear basis to judge one type of error or the other (i.e. errors that favor the plaintiff or the defendant) to be worse, so civil suits are supposed to be decided without bias, on the basis of "the preponderance of the evidence." In matters of foreign policy and national security, US policy often favors extremely costly action to defend against threats that are not just uncertain but unlikely, because the cost of being unprepared to meet a threat that does materialize is judged to be so severe.[14]

In any policy area, it is possible to bias decisions either for or against action – environmental activists often use the same argument for a pro-action bias as is made for national security – but this choice is not scientific; rather, it reflects a judgment about what errors we want to avoid. What approach to decision-making is appropriate for climate change? The argument that climate science is too uncertain to merit action, and the analogies to criminal law and scientific research on which it is based, would reject any mitigation actions until highly confident projections of severe climate-change impacts were available. This is a difficult standard to achieve: such confidence might never be achieved until the impacts were already realized or too late to avoid. But this approach would be appropriate if it was judged much worse to limit emissions too much than not to limit them enough – i.e. that the economic losses from too much mitigation were much worse than the impacts of too much climate change. There is no basis for thinking this to be the case, however; rather, the reverse situation appears more likely. If uncontrolled climate change and its impacts turn out to lie at or below the bottom of the present projected range, then an aggressive mitigation program would impose substantial unnecessary costs, presently estimated to lie between a few tenths of a percent and several percent loss of future GDP. But if climate change and impacts lie near or above the top of the present projected range, then not pursuing aggressive mitigation would likely expose the world's people to much more severe costs and risks, including a growing possibility of abrupt, perhaps catastrophic changes.

With high and uncertain stakes on both sides, a response to climate change requires decisions under uncertainty that consider risks and potential costs symmetrically, acknowledging the risks both of responding too strongly and not responding strongly enough. In this respect, climate change resembles all other first-rank policy issues, including responding to security threats such as hostile foreign powers or terrorism, making economic policy, and managing all kinds of

---

[14] For example, Secretary of State (at the time) Colin Powell made the following statement of why the USA was pursuing national missile defense: "[T]here is recognition that there is a threat out there . . . And it would be irresponsible for the United States, as a nation with the capability to do something about such a threat, not to do something about [it] . . . you don't wait until they are pointed at your heart. You start working on it now." (Remarks at the International Media Center, Budapest, Hungary, May 29, 2001, http://www.state.gov/secretary/rm/2001/index.cfm?docid=3126)

risks to life, health, and safety. This stance requires rejecting the argument that the mere presence of uncertainty requires delay, since waiting for near-certainty could carry high costs or take forever. It equally requires rejecting the opposite extreme stance, that climate change is a crisis demanding the maximum possible response immediately, regardless of cost or consequences. Whether this latter stance is framed as the strongest form of the precautionary principle, or by characterizing greenhouse-gas emissions as a moral wrong that must be eliminated, it is as insupportable as the stance that no mitigation should be made until decisive evidence compels it.

Unfortunately, an approach balancing risks on both sides does not gain much resonance with either side in the present, ideologically charged debate. But however unpopular this approach may be, it is essential. While there is much uncertainty about the consequences, costs, and benefits of alternative courses of action, we must consider the knowledge we do have in choosing among actions – which is sufficient to reject certain extreme courses of action, even if it is not sufficient to specify the precise course we should follow.

5.4   So what should be done? Major choices and elements of an effective response

Our judgment is that present knowledge and evidence of the risks of climate change are sufficient to demand strong action, despite continuing uncertainties of varying magnitude and significance in nearly every aspect of the issue. Given the risk of serious, slow-to-reverse harms, it would be irresponsible to wait for precise knowledge of the form and magnitude of climate-change risks before taking action to forestall the risks. The response to climate change must reflect uncertainty, of course. But this means balancing the risks of acting too strongly and acting not strongly enough, and maintaining the flexibility to adjust responses over time as knowledge advances and conditions change – not waiting for certainty before taking action.

What form should the response take? We have already laid out some of the obvious and uncontroversial elements of a response. Continued scientific research on climate and its impacts is essential, but not enough. Advance planning and building adaptive capacity, to prepare for a more uncertain and probably less benign climate than we have experienced for the past century are essential, but not enough. Supporting independent, high-quality scientific assessments to inform continuing policy decisions is essential, but not enough. In addition to these elements, an effective response to climate change must also include a strategy to reduce global emissions, starting soon and continuing for decades until most of the world's energy system has shifted to non-emitting alternatives. Such a mitigation strategy comprises four elements: a long-term target that, based on present knowledge, appears to adequately protect the global climate; feasible, well designed near-term policy initiatives to move toward the long-term goal; a political strategy that motivates participation in the required near-term actions and that is consistent with advancing toward the long-term goal; and a mechanism for adapting both goals and actions in light of evolving knowledge, experience, and capabilities.

5.4.1   Long-term goals

A climate-change mitigation strategy can benefit in several ways from having an explicit long-term goal. A goal that is challenging but attainable can focus attention, motivate action, and provide a context for choosing and evaluating near-term measures, even if the relationship between the near-term measures and the long-term goal is known only approximately. The long-term goal stated by the Framework Convention is stabilizing atmospheric greenhouse-gas concentration "at a level that would prevent dangerous anthropogenic interference with the climate system." This is a fine goal, which gains nearly universal agreement when stated at this level of abstraction, but because it depends on defining how much interference is judged to be "dangerous," it is too vague to be operational. Using global-average temperature change as the measure of disruption, proposed warming limits to avoid dangerous interference have ranged from 1 °C to 5 °C, with most proposals lying between 2 °C and 3.5 °C. (Recall that a change of 3 °C is five times the warming realized over the twentieth century. If this occurred by 2100, as present projections suggest is likely, this would represent double the rapid warming rate of 1970–2000, with this doubled rate sustained for 100 years.)

There is no bright line that demarcates dangerous interference with the climate system, of course. This is so both because of uncertainty about the impacts of different levels of greenhouse gases, and because of disagreement about what impacts are acceptable and what efforts are worth making to avoid them. There are neither precise moral principles nor decisive practical considerations that can tell whether the proper temperature limit is 2 °C, 3 °C, or 4 °C, or some higher or lower value. But waiting for either definitive benefit–cost analysis or complete political consensus to identify a precise goal – just like waiting for elimination of scientific uncertainty – would mean waiting forever – or at least so long that most desirable goals would have long become unattainable. In our judgment, limiting total global warming to 3 °C is an appropriate goal.[15] Present knowledge suggests that this limit is achievable with sustained efforts that are serious, but

---

[15] Note that this warming target is defined relative to temperatures before the warming of the twentieth century, so this target means about 2.4 °C additional warming above the present global temperature.

Table 5.1. *Limits for atmospheric concentration of all greenhouse gases (expressed as CO₂-equivalent in p.p.m.), for selected combinations of doubled-CO₂ climate sensitivity (an uncertain property of the climate system) and limits on global warming (a choice)*

| | | Doubled-CO$_2$ climate sensitivity (°C) | | |
		1.5	3	4.5
Warming limit (°C)	2	710	440	380
	3	1120	560	440
	4	1780	710	520

Source: calculated from Caldeira *et al.* (2003).

not overwhelming and would likely avoid the most severe risks – although it still represents extreme climate change relative to the experience of human civilization, and some scientists dispute that it would avoid severe risks.

Given this or any other specific limit on long-term global warming, and an assumed value for climate sensitivity, it is possible to infer a limit on atmospheric greenhouse-gas concentrations. If climate sensitivity is high, then limiting warming to any specified level requires limiting greenhouse gases to lower concentrations than if sensitivity is low. Suppose, for example, that doubled-CO$_2$ climate sensitivity is 3 °C, the middle of the estimated range. Then limiting future warming to 3 °C requires limiting greenhouse-gas concentrations to about 560 p.p.m. of CO$_2$-equivalent. Relaxing the warming limit to 4 °C (still assuming sensitivity of 3 °C) would let concentrations increase to about 710 p.p.m. of CO$_2$-equivalent, while tightening the warming limit to 2 °C would require limiting concentrations to about 440 p.p.m. Table 5.1 summarizes how the required limit on greenhouse-gas concentrations depends on the combination of the warming limit and the climate sensitivity.

These limits apply to the total climate-forcing effect of all greenhouse gases, expressed as an equivalent concentration of CO$_2$. Since this includes the effects of other greenhouse gases, CO$_2$ itself must be stabilized at a lower level. How much lower depends on how fast emissions of the other gases grow. Table 5.2 shows approximate implied concentration limits for CO$_2$ itself, given mid-range assumptions for growth of other gases.

These limits paint a sobering picture of our situation. They suggest that achieving any but very weak limits on global climate change (i.e. limits that allow very substantial risks), under relatively fortunate assumptions about the world (i.e. low to middle climate sensitivity) will require a massive deflection of present emission growth trends. These numbers also indicate how important it is for a mitigation

Table 5.2. *Approximate limits for atmospheric concentration of CO$_2$ alone (in p.p.m.), for selected combinations of climate sensitivity and limits on global warming, assuming a mid-range growth path for non-CO$_2$ greenhouse gases*

		Doubled-CO$_2$ climate sensitivity (° C)		
		1.5	3	4.5
Warming limit (°C)	2	510	320	270
	3	810	400	320
	4	1280	510	370

Source: interpolated from Wigley, Stabilization of greenhouse-gas concentrations, in Aspen Institute (2002).

strategy to include limits on non-CO$_2$ gases, to make the CO$_2$ part of the problem a little less overwhelming.[16]

Taking our proposed goal of limiting global warming to 3 °C, and assuming that climate sensitivity is 3 °C, roughly in the middle of the presently estimated range, atmospheric concentration of CO$_2$ must be stabilized around 400 p.p.m., or somewhat higher if substantial reductions in emissions of other greenhouse gases can be achieved. Assuming this is the case, we will examine the implications of stabilizing CO$_2$ around 450 p.p.m.

The present concentration and its rate of change – 380 p.p.m., increasing by 2 p.p.m. per year – give an indication of how challenging it will be to keep atmospheric CO$_2$ below 450 p.p.m. Indeed, some analysts of energy and emission trends regard 450 p.p.m. as already out of reach, and treat stabilizing at 550 p.p.m. – double the pre-industrial concentration – as the lowest level that is technologically, economically, and politically feasible. Yet stabilizing at 450 p.p.m. would still expose us to significant risks. One recent attempt to quantify risks of dangerous climate change found that even 450 p.p.m. was associated with a 35 to 40 percent chance of surpassing the threshold of danger. Aggressive non-CO$_2$ reductions and adaptation measures reduced this risk, but only to 15 to

---

[16] As we discussed in Chapter 3, trends in atmospheric aerosols will also influence how hard it is to limit future climate change. But because of uncertainties in both present effects and future trends in aerosols, their aggregate effect is uncertain even in its direction. Recall that SO$_2$ emissions produce a net cooling effect. Consequently, if their present total effect is large and they decrease rapidly in the future, then even more stringent limits on CO$_2$ and other greenhouse gases will be required than shown in Tables 5.1 and 5.2. On the other hand, black aerosols produce a net warming effect. Consequently, if their present total effect is large and they decrease rapidly in the future, then the required limits on CO$_2$ and other greenhouse gases will be somewhat less stringent than shown in Tables 5.1 and 5.2.

20 percent. Although this was a preliminary and illustrative analysis, it suggests that it might be premature, indeed irresponsible, to abandon the prospect of limiting atmospheric CO_2 to 450 p.p.m.

5.4.2   Near-term actions

What does the long-term goal of stabilizing CO_2 around 450 p.p.m. mean for emission trends and required actions in the near term? As Figure 4.3 showed, low-cost emission trajectories to reach this goal require that global CO_2 emissions begin deflecting from their present growth path in just a few years, peaking slightly below 10 GtC/yr around 2010 and then declining to about 6 GtC/yr by 2050 and less than 4 GtC/yr by 2100. It is possible to reach the same stabilization level with a somewhat later start to emissions' divergence from their present growth path, but only if their subsequent decline is substantially faster. Relaxing the concentration-stabilization goal to 550 p.p.m. would let global emissions continue to grow to about 12 GtC around 2030 before turning downward, declining to about 6 GtC by 2100. But adopting this weaker goal would mean either gambling that climate sensitivity lies near the bottom of its estimated range, or accepting a global temperature rise of more than 4 °C. If baseline emissions lie near the middle of the projected range (as shown in Figure 3.11), they will increase by about 1.5 GtC per decade over the next few decades. Consequently, emission trajectories that aim to stabilize at 450 p.p.m. will require global emissions to be reduced by about 1 GtC below the projected baseline by 2015, 5 GtC below the baseline by 2030, and 10 GtC by 2050, by some combination of increased efficiency and switching to energy sources that emit no CO_2 to the atmosphere. Considering the projected sharp growth of developing-country emissions over this period, industrialized-country reductions must be much larger, for example to about 60 percent below 2000 levels by 2050 if developing countries begin to control their emissions only around 2030.

This estimate of required reductions depends on several points that are either uncertain or matters for choice: the target for maximum global warming (a choice); the climate sensitivity (an uncertainty); the baseline emissions trend (an uncertainty); and the trajectory of non-CO_2 emissions. We cannot at present either identify a precise choice of climate-change limit, or eliminate these uncertainties. But this is of little importance for the choice of near-term actions, because what is required in the near term is similar for a wide range of targets and assumptions. Anything except a combination of weak climate-change targets and favorable assumptions will require substantial downward deflection of global emission trends starting within 10–20 years. Given how long it takes to develop and implement policies, develop and deploy technologies, write off investments, and change behavior, this means that development of effective policies and technologies to reduce all greenhouse gases, especially energy-related CO_2 emissions, must begin immediately to avoid increasing costs and risks of failure.

We have stressed the crucial role of a large increase in government spending on research and development of multiple energy-related technologies. But while such public effort is essential, the success of a mitigation program will stand or fall on how well it motivates private-sector efforts to change present products and production processes and to deploy the R&D and investments needed to bring these changes about. Mobilizing these efforts will require a strong, credible public-policy signal that emitting greenhouse gases will grow increasingly costly over the next few decades. Voluntary and information-based programs may complement and enhance core policies around the edges, but they cannot achieve the required changes. The present reliance of climate-change policy on voluntary programs, in the USA and elsewhere, is woefully inadequate in view of the severity of the challenge – as much as it would be to rely on voluntary actions to finance the government or provide for the national defense. Only binding, authoritative policies that carry real incentives can provide the structure, clarity, planning environment, stability, incentives, and leadership that are required to motivate the required changes in private decision-making, principally by business.

What form should these policies take? Many aspects of the required policies have been widely discussed and widely agreed. Policies should seek to minimize costs by allowing flexibility in implementation, through harnessing market forces to the extent feasible. They should be announced well in advance and phased in gradually, to limit costs and allow stability for planning. Beyond these agreed elements, commentators differ on whether the preferred form of mitigation policies should be national emission limits, carbon taxes, conventional regulations targeting performance in specific sectors, or some combination of these. Different forms of policy may be appropriate in different nations, even while mitigation efforts must eventually be coordinated globally. In view of present uncertainty about mitigation costs, and the risk of backlash if early costs rise too high, our view is that the preferred policy is a tradable emission permit system including an escape valve – a commitment to sell additional permits if their price rises above some specified level. For the USA, a suitable initial level for the emission cap might be somewhat tighter than the McCain–Lieberman bill but somewhat weaker than the Kyoto commitments – for example, 5 to 7 percent above 1990 emission levels in 2010 (rather than 6 percent below as in Kyoto), coupled with a pre-announced trajectory of further cuts that gradually increase in stringency. A similar target might be most suitable in the near term for those nations that face the greatest difficulties in meeting their Kyoto targets, such as Japan and Canada – although these nations are of course committed to meeting substantially stricter targets.

The purpose of the escape valve is to limit economic harms from unexpectedly high costs that might arise if the emission limit is tightened too fast, by putting a ceiling on marginal cost. To keep the emission limit meaningful, however, the escape-valve price should be set high enough that it is relatively unlikely to be reached: a suitable initial value might be $75 to $100 per ton of carbon, equivalent to about 18 to 24 cents per gallon of gasoline or 0.8 to 2.4 cents/KwH of electricity, depending on the fuel source. To reduce the total cost burden and help smooth out energy price fluctuations, the escape-valve price could be decreased in parallel with fuel price increases above a certain level, so permit prices would be allowed to rise higher when fuel is cheap than when it is expensive.

Although the most efficient way to distribute permits initially would be by auction, securing enough support to establish the scheme would probably require that some fraction of the permits be distributed free of charge to present emitters. While including the broadest possible collection of greenhouse gases, economic sectors, and activities within the trading system would reduce the cost of achieving the reductions, this advantage of a broader emission-trading system must be balanced by practical concerns about how well it is possible to monitor and account for emissions. A system that includes only energy-related CO_2 emissions is not ideal, but may be all that can be practically implemented as a first step. As the ability to monitor other emissions advances, the scope of the permit system should be broadened to include additional emission sources and gases.

The tradable-permit system and escape valve would provide the central component of a mitigation policy, but other policies could complement them. In particular, there would be a role for additional regulatory policies, either market-based or conventional, in areas where two conditions apply: the technical potential for emission reductions at relatively low cost appears to be large; and there is reason to doubt the effectiveness of energy-market price signals induced by emission limits. Regulation of vehicle fuel economy is a prime example, because the technical efficiency characteristics of cars and light trucks have a large impact on overall emissions but respond only weakly and slowly to changes in fuel prices. Stronger policies could motivate substantial improvements in vehicle efficiency, although ideally this should be achieved through a new policy more efficient than the present, conventional regulatory system, known as the CAFE standard. For example, a system of tradable permits or fees could be applied to the standardized fuel consumption of newly manufactured or imported vehicles, both automobiles and light-duty trucks. Another promising example would be abandoning the long-standing preferential treatment on air pollution from old power plants under the US Clean Air Act. This policy has the effect of keeping inefficient old plants, which are far worse than new plants in their emissions of greenhouse gases as well as conventional air pollutants, in service long after their expected lifetime.

How much will these mitigation policies cost, and are they worth it? Opponents of mitigation assert that costs will be high, perhaps ruinous, so only very limited efforts are warranted, at least for the near term. But the safety valve would eliminate the risk of the highest projected costs in the event that mitigation turned out to be unexpectedly difficult and expensive. Moreover, as we discussed in Chapter 4, how sharply we would wish to cut emissions depends not just on the cost of mitigation, but on the balance of costs of climate impacts, mitigation, and adaptation, all of which are quite uncertain. While climate impacts will likely be modest for rich countries in the near term if climate change lies near or below the middle of the range of present projections, the principal concern with climate change is not these impacts, but the possibility of far more serious impacts from changes near the high end of present projections, or from mid-range changes sustained beyond 2100, or from potential abrupt changes that present projections do not consider. These higher impact scenarios, which are virtually ignored in present climate-impact assessments, carry a small, non-negligible probability of very serious harms – an uncertainty that cannot be eliminated until the actual changes, whether severe, modest, or in between, are upon us and cannot be reversed for many decades or longer.

Mitigation costs also carry uncertainty, but the origin and implications of this uncertainty are quite different from uncertainty in climate impacts. The substantial uncertainty in mitigation costs that comes from different policy assumptions simply provides guidance for how policies should be designed to keep costs as low as possible. The remaining uncertainty – which mostly concerns how much technological innovation can reduce the cost of cutting emissions – can, like climate-impact uncertainty, best be reduced through experience. In the case of mitigation, this experience in part will mean actual efforts to develop and deploy new low-emitting technologies. Motivating the required efforts, which must principally come from the private sector, will require policies that generate strong enough incentives, and that send credible signals that the incentives will not be removed next year.

We cannot know the results of such efforts until we make them, but in general they are likely to lead to identification of more opportunities to reduce emissions and reduced estimates of the cost of doing so. Experience from other environmental issues suggests that costs for reducing most forms of pollution turn out to be lower than they are projected to be in advance. Moreover, the highest present mitigation cost estimates make assumptions about technological response to incentives and the ease of substitution in the economy that are about as pessimistic as could plausibly be true. Consequently, in contrast to climate-change impact projections, actual mitigation costs are quite unlikely to lie well above the present range of estimates. There is one important qualification to this claim: inefficient, badly designed policies could drive mitigation costs far above what they need be. This is a serious political risk that cannot be ignored. But except for this risk, the process of resolving uncertainties and learning more about mitigation is likely to push the present range of projected mitigation costs down, not up. Moreover, even if mitigation costs decline only slowly, any harm suffered from reducing emissions too fast is likely to be more readily reversible than the harm from allowing too much climate change to happen, particularly if the mitigation policy includes an escape valve or other provisions to slow emission cuts in the face of persistently high costs.

5.4.3   A political strategy

At present, the world is far away from having any mitigation regime that could make a serious contribution to limiting climate change. A few countries are developing serious mitigation policies, but even these fall well short of what will be needed to achieve the required shifts in emissions. Most nations have policies vastly too weak for the job, or none at all. The essential political problem of managing global climate change is to identify a series of achievable steps to move progressively from this present state, toward the goal of widespread adoption of serious, cost-effective, coordinated mitigation policies. Because nations cannot be coerced to join an international climate-change regime, a feasible political strategy will require deploying incentives that will lead the required participants to join the regime, and to meet their obligations under it, voluntarily.

An effective international mitigation strategy must be able to produce the required large-scale reduction in global emissions, and transformation of the world energy system, over the next several decades. Achieving this requires a mitigation strategy to satisfy several criteria. It must include a feasible first step that can break the present deadlock. Early steps must promote, not hinder, subsequent movement toward progressive reductions of emissions and expansion of participation. Because expanding mitigation opportunities requires large increases in R&D and investment in energy technologies, the strategy must support and motivate these investments. Because the costs of greenhouse-gas mitigation are likely to be substantial, the strategy must be cost-effective – i.e. it must be structured to achieve the required energy-system changes and global emission reductions as cheaply as possible, over the entire relevant time horizon. The strategy and its implementation must also provide adequate incentives for both governments and private actors to participate, and to make good-faith efforts to meet their commitments.

Finally, the strategy must be sufficiently equitable in its distribution of burdens to gather widespread support. In particular, it must reflect some defensible interpretation of the principle of common but differentiated responsibility, so developing-country burdens reflect their different status and do not obstruct their development. This is both a normative requirement and a practical one, since starkly inequitable approaches are unlikely to gain the widespread support and legitimacy necessary to motivate participation and good-faith performance. Note, however, that this is a substantially weaker condition than often proposed in climate policy debates. It does not require explicit negotiation of equity criteria or burden-sharing formulas. Nor does it require that the climate-change regime make a large contribution to redressing present global inequities.

In choosing an international mitigation strategy in light of these criteria, the most basic near-term political choice is whether to stay with the present structure, as embodied in the Framework Convention and the Kyoto Protocol, or to make large-scale departures in the architecture of agreements or the set of participants.

The Kyoto Protocol is in an awkward situation. Rather than dying, as its opponents hoped, it has entered into force. But it has several fundamental problems, on which little progress has been made for several years. First, no industrialized country appears likely to achieve large enough domestic emission reductions to meet its first-round emission obligations except the EU, and even they may fall slightly short. Parties may formally meet their commitments by buying surplus credits, of course, but few would regard large-scale reliance on this means of compliance as success – and the prospect of sending large checks to Russia may provoke substantial opposition to complying via this route. In addition, parties have made very limited progress in negotiating further emission cuts after 2012; no progress in engaging the USA, despite increasing indications that other parties are willing to offer almost any adjustment of the current US commitment; and no progress at all in engaging developing countries in serious negotiations about their future mitigation commitments.

Many elements of the Protocol can in principle be renegotiated: the form and level of mitigation commitments, of course, but also the mechanisms for implementation, reporting, and compliance; the terms of the flexibility mechanisms; or what emissions and activities are included and how they are counted and converted. Other aspects of the current approach are more firmly embedded, such as the basic approach of controlling emissions via quantified national emission targets, the decision to allow various forms of flexibility in meeting these targets, and strong differentiation of commitments between industrialized and developing countries. This differentiation is firmly established in both the Protocol and the Framework Convention. Indeed, although many proposals are being made to extend mitigation commitments to developing countries – for example by nations "graduating" to stricter targets as they pass pre-agreed GDP thresholds – the listing of particular nations in Annexes 1 and 2 of the Convention makes such expansion awkward to negotiate,[17] and there is little indication that developing countries are willing to consider such approaches.

The most basic structural aspect of the present approach, however, is universal participation. Both the FCCC and Kyoto Protocol have sought the broadest possible participation from the outset. Virtually all nations of the world participate in climate-change negotiations (as of April 2005, the Framework Convention has 194 parties, the Kyoto Protocol 148), and all the rich industrialized countries and many former Soviet states and allies initially agreed to emission limits in the Protocol's first commitment period. Universal participation has been pursued both because it is expected to lower costs and because it is viewed as more legitimate. Broad participation in mitigation is expected to lower costs through the Protocol's flexibility mechanisms, because they allow international shifting of mitigation effort to where it is cheapest – in the first commitment period, principally by buying unused emission credits from Russia. Broad participation in negotiations is also valued because climate change and responses to it have the potential to transform many areas of national policy and international relations. With such stakes, all nations reasonably wish to be involved in the early negotiations that might shape the subsequent direction of the regime.

Proposals to escape the present deadlock fall into three broad categories: some propose to keep all major elements of the Protocol but negotiate specific, relatively minor changes to resolve the present problems; others propose more substantial revisions to the architecture of the Protocol and/or Convention – for example changing the form of mitigation commitments – while still working within these instruments and retaining universal participation for all negotiations. The most radical proposals would abandon universal participation – at least as a transitional stage – and seek other vehicles for international cooperation involving smaller groups of nations. We briefly consider each of these.

Many specific, relatively minor changes to the Protocol have been considered. These proposals all retain universal participation, national emission targets with flexibility mechanisms, and differentiation of commitments. Some of them focus on particular improvements to the implementation system or flexibility mechanisms, or propose tuning the mitigation commitments to more closely follow a low-cost path toward some concentration stabilization level. Others focus on specific changes intended to persuade the USA and the developing countries to accept mitigation commitments: for example, for the USA, a looser target in the first

---

[17] Changing the countries on these lists, or replacing the lists with criteria or procedures to determine who is in each group, would require amending the Convention. Amendments require the support of at least three-quarters of the parties and, even if adopted, only become binding on parties that formally accept them via a process equivalent to ratification of the original treaty.

commitment period, a different baseline year, or a large one-time credit for sinks; and for the developing countries, a promise to reinvigorate the Clean Development Mechanism (a mechanism to finance emission reductions in developing countries), or the offer of emission targets that follow or even exceed their projected baseline emissions in the near term, to reduce their risk and let them sell extra credits. These generous developing-country targets are intended to hold them harmless, or even to let them profit from joining the mitigation regime, at least for the first few decades.

The difficulties with these proposals are clear from the very fact that they have been thoroughly circulated and discussed over the past few years, with no progress. Persuading the USA and the developing countries to join by such modest changes is not impossible, of course. But it would require a large departure from present positions, and larger-scale political postures, on the part of both. US opposition to the Protocol goes deeper than objections to the first-round US target. It includes significant opposition to the structure of commitments, and the expectation that any commitments now being proposed, even in future periods, would differentially disadvantage the USA – principally because of the higher pre-existing energy taxes in other industrialized countries. Similarly, developing-country resistance to accepting mitigation commitments goes beyond simply wanting favorable allowances. In addition, their opposition reflects both the principled view that industrialized countries must show real efforts and progress before developing countries are asked to follow, and the suspicion that with large uncertainty in future emission growth, even seemingly generous initial allocations might constrain development, and that if they accept the principle of developing-country emission limits they would risk being pressured in subsequent negotiations to accept much tighter limits that would erode any advantage they held and shift the burden to their disadvantage.

In the most hopeful scenario, the USA and the developing countries might each be willing to join some form of mitigation commitments if the other does – i.e. there are no concessions by present parties that could bring them in separately, but they could be brought in together. The role of the present Kyoto parties would be to broker a deal between the USA and the developing countries for their joint accession, with whatever modifications to the Protocol are necessary to let this happen. But if there are potential mutually advantageous agreements to be found between the USA and the developing countries, there is no clear advantage to them in letting the present parties play that role. They would surely prefer to do it on their own, outside the framework of the Kyoto Protocol, and keep control of the negotiating agenda between themselves. We discuss this prospect below.

Alternatively, many commentators have proposed more far-reaching revisions to the architecture of the Kyoto Protocol that still keep the fundamental elements of differentiated commitments and universality. These proposals are numerous and diverse. They include, for example, more complex trading systems with permits of multiple durations to help manage risk; adding an internationally established escape valve to the international trading system; permits whose initial distribution is so abundant that they do not constrain emissions, but some of which are subsequently repurchased and retired by an internationally financed authority; letting developing countries voluntarily accept high pseudo-baselines that represent not a regulatory requirement, but only an accounting point below which they are allowed to sell permits; shifting negotiations away from national emission limits toward mutually agreed actions, such as a common carbon tax, support and incentives for R&D, technology-based standards for major emitting sectors, or some broad collections of policies and financial and technical assistance.

Each of these proposals responds to one or more identified problems with the present approach, and these proposals hold somewhat more promise to break the current deadlock than the modest revisions discussed above. But all these approaches still suffer from fundamental weaknesses that are related to universal participation and to commitments so strongly differentiated that many nations have none at all. Universal participation in negotiations, together with a norm of decision-making by consensus, creates powerful opportunities for obstruction. A group of nations who are willing to take on some mitigation commitments cannot negotiate their terms, design, or implementation without many other nations who are not accepting the commitments having a voice, or even a veto.

This mismatch between who is undertaking efforts and who is negotiating the terms of the efforts obstructs effective negotiation in many ways. It can separate negotiations from considerations of practicality, since most negotiating parties will not have to do what they are discussing. It can obstruct attempts to negotiate changes in the set of nations that have mitigation obligations, or to develop incentives to motivate additional nations to accept them. It can allow negotiation of initial mitigation commitments to become a vehicle for non-participating nations to secure favorable precedents and maneuver for long-term advantage. Most seriously, universality empowers some states, principally major fossil-fuel exporters, who oppose any attempt to establish a mitigation regime. Because these nations seek not just to avoid mitigation commitments themselves, but also to prevent others from adopting them, their primary objective in negotiations is to obstruct progress. While the universality of the Convention and Protocol means that these nations cannot be excluded from mitigation negotiations, the norm of consensus decision-making means that their obstructive tactics are frequently effective.

In view of the clear obstacles to negotiating a mitigation regime through any universal process, some commentators have proposed moving outside the Kyoto framework. The most prominent suggestions include building international permit markets before negotiating emission limits; negotiating a bilateral deal between the USA and China; and negotiating stronger commitments among a relatively small group of industrialized nations most committed to establishing a mitigation regime.

One proposal would shift the focus of initial international activity away from negotiating emission targets and toward first constructing a well-managed international market in emission permits (see, for example, "proposal 3" in Victor, 2004). Several emission-trading systems are already in operation. The EU's system is the most developed, with compulsory participation for major stationary emission sources, substantial penalties for emitters who exceed their allowance, and a requirement for annual emission reductions. Substantial trading systems are also in operation inside several major multinational corporations, and among the 75 major US industrial firms who are members of the Chicago Climate Exchange. This proposal would move toward global emission control by progressively strengthening, linking, and expanding emission-trading systems, thereby allowing exchanges over a progressively wider set of regions and activities. The proposal would seek to develop permits to emit a unit of greenhouse gases into a new global financial instrument – albeit, one that would only influence emissions if every participating nation took steps to ensure the integrity of the instrument. For example, every participating nation would have to set a limit on the number of permits it issues, monitor emissions within its territory, and effectively enforce the requirement that emitters hold a permit. Someone – perhaps the largest sponsor nations or the exchanges where permits are traded – would also have to assess how well each participating nation meets these requirements, and also how reliably the reductions from various activities can be counted and verified. For example, sequestration in forests and soils is potentially a large sink for CO_2, but is both hard to measure and at risk of returning to the atmosphere early if environmental or economic conditions change. Liability for accounting errors and project failure would need to be assigned, most likely to the buyers of permits. Under this condition, permits generated by activities that are hard to monitor and ensure, and those issued by countries with lax monitoring and enforcement, would be expected to trade at a discount relative to permits based on clear, secure, and well enforced reductions.

The fundamental problem with this approach – at least in the pure form that completely separates establishment of an emission permit market from negotiation of national emission limits – is that it leaves the decision of how many permits to issue up to each participating nation. This would make the collective-action problem among nations even more severe than in explicit negotiations over national emission limits. Nations issuing excess permits do not merely get to free-ride on leaders' efforts, or gain some inflow of emissions-intensive investment; they can thwart leaders' attempts to reduce emissions even in their own territory, because permits issued by countries issuing many can be sold to emitters in nations issuing fewer. Avoiding this situation would require negotiations to limit how many permits each nation issues, which would be as difficult and as unlikely to achieve significant reductions as current negotiations over national emission limits. But in the absence of agreed binding limits, permits would be so abundant that their value would be near zero – except to the extent that, like the present US voluntary permit market, they price the risk of more stringent controls being imposed in the future. With permits nearly worthless, nations and permit-holders would have little incentive to invest in the monitoring and enforcement that will be necessary to defend the integrity of a permit system when values grow higher. In view of all these difficulties, it is unlikely that the prior creation of an international permit market can avoid the hard negotiations and need for political will that are necessary to impose non-trivial limits on emissions.

Moreover, a central thrust of any proposal to establish a permit market as an *alternative to* negotiating national emission reductions – rather than as a *means to efficiently implement* a negotiated agreement on national emission reductions – is to fix national baselines, and consequently the starting point for any future mitigation negotiations, at present levels. By enshrining the status quo, this approach consequently avoids any consideration of equitable global distribution of emissions. While we have criticized the universal approach of the Kyoto Protocol as encouraging too much preoccupation with broad questions of global equity, a system that goes no further than solidifying present inequities would in our view have little hope of gaining the widespread participation necessary to make any significant contribution to limiting global emissions.

A second approach that moves outside the Kyoto framework would be bilateral negotiations between the USA and one of the largest developing country emitters – in most proposals, China (see, for example, Stewart and Wiener, 2003). These bilateral negotiations would establish a firm baseline emission trajectory for each nation, a reliable system to account for emissions, and a mechanism to exchange emissions either at national or project level. Pursuing cost-effectiveness would probably require that the bulk of near-term mitigation activity take place in China rather than the USA. Because this mitigation would in all likelihood carry a cost (even after accounting for the higher efficiency of new capital equipment, and the prospect of co-benefits such as air pollution reduction and energy supply security), the arrangement would have to include adequate incentives for China to participate, probably in the form of enough excess emission permits that the revenue they can expect from selling them will offset the costs they incur from mitigation, leaving them at least as well off as under their baseline emission growth projection.

By engaging China in a market that puts a significant price on emissions immediately, this proposal holds the promise of influencing the rapid buildup of capital investment now underway there – motivating additional investments to improve the efficiency and reduce the associated emissions of the new capital stock, and so gaining emission-reducing benefits for decades.

Negotiating a comprehensive, well implemented trading system bilaterally would also have the advantage of being vastly simpler than attempting to negotiate such a system in the global context of the Kyoto Protocol. It could allow the orderly construction of a permit market initially at a manageable scale, with the possibility of experimentation and revision of specific institutional details. It could also provide a prominent model of a deal between industrialized and developing countries that reduces emissions and is advantageous to both parties. Once established bilaterally, the arrangements could be expanded through voluntary accession to the system by additional countries, both industrialized and developing, thereby expanding the scope of the trading system and the associated opportunities for cost savings. Although the initial development of this approach would proceed entirely outside the Protocol, the aim would be eventually to negotiate a merger of the systems.

In our view, this approach holds more promise than the previous suggestions. It would of course require much more serious engagement from both the USA and China than either has demonstrated so far: in effect, it presumes that the unwillingness of each of them to participate can be overcome by securing the participation of the other. This might be so – or might come to be so, given imaginable political shifts in each country. The most serious question about the viability of this proposal concerns the size of financial transfers implied by allocating excess permits to China, whether these come from public sources or from firms paying to avoid domestic mitigation obligations. In the present budgetary and political context, it is not clear whether the USA would be willing or able to pay these costs to secure Chinese participation in mitigation.

There is one more potential path leading from the present deadlock toward a viable global mitigation regime that in our view is more promising than any discussed so far: development of a serious mitigation strategy by negotiations among a relatively small group of similarly situated, rich industrialized countries – a "coalition of the willing." The best candidates for this coalition are those nations that have demonstrated the most serious commitment to mitigation thus far – principally the European Union, perhaps with additional industrialized-country parties to the Protocol. These countries would negotiate initial agreements on a long-term atmospheric target, and progressively more stringent near and medium-term national mitigation obligations consistent with that target, with accompanying measures to limit the associated risks.

There are three types of risk associated with this approach that must be limited: that the participating coalition might be ineffective because it is too small to deploy the incentives required to shift global private-sector R&D and investment; that participants might suffer competitive disadvantages so severe that the approach is politically unsustainable; and that the approach might obstruct subsequent expansion toward near-global participation in a mitigation regime.

We have argued above that effective mitigation policies must be globally coordinated *eventually*, but full global participation is not necessary immediately. But how many are enough: which nations must join at the start for a mitigation regime to be effective, politically sustainable, and consistent with movement toward global participation? The first condition, effectiveness, depends principally on the total size of the participating economies. The participating nations must make up a large enough market that their policies can influence the research, investment, and operational decisions of both domestic and foreign firms to reduce emissions from their products and production processes. They must also be big enough collectively to deploy incentives that other nations will take seriously in formulating their own policies, and to attract enough attention to shape the agenda for subsequent international negotiations. A hint about the required scale of initial participation is provided by the several decades of success of California at independently regulating pollution from automobiles and driving technological advances that have improved emission performance worldwide. In view of the success of this small jurisdiction – about 3 to 4 percent of the world's economy and its market for automobiles – it is highly likely that either the EU or the United States, or any broader coalition of rich industrialized countries, would be big enough to take the first step in establishing an effective mitigation regime.

We discussed above the practical and principled reasons to favor broad participation in mitigation. These reasons are valid and important, but the present approach based on universal participation has not brought any significant progress toward an effective international mitigation regime. We contend that the practical advantages of starting with narrower participation may be so great that they outweigh the advantages of breadth.

Indeed, the importance of the cost advantage from breadth and international flexibility in the Kyoto Protocol may be over-rated. In the case of acquiring excess credits from Russia, this cost saving is essentially fictitious: to the substantial extent that these acquired credits represent Russian emissions that would not have occurred in any case, these exchanges lower the cost of emission reductions by not reducing emissions. But even where real reduction effort is moved abroad rather than avoided entirely, the cost advantage mainly comes from exploiting cheap opportunities to replace or upgrade old and inefficient equipment using newer and more efficient – but presently available – technology. This is clearly worth doing, and the associated shifting of effort does reduce costs in the short run. But the opportunity to shift reduces the stringency of incentives to reduce, and to develop lower-emitting technologies, that are felt by firms and countries that face high marginal costs – who in many cases will also be those with best access to the financing and technological capability to pursue these innovations. Short-term cost minimization may consequently come at the cost of weakening incentives to develop the new technologies required to reduce the cost of larger, long-term reductions, and may thus serve to delay the development of needed capabilities to resolve the climate issue.

Narrower initial participation in the mitigation regime would sacrifice some of this near-term cost-saving opportunity but deploy stronger incentives to develop new, non-emitting technologies capable of making larger contributions to long-term emission reduction. Limiting initial participation to a group of willing, similarly situated, rich industrial countries would greatly limit incentives and opportunities for procedural obstruction, and would allow negotiations to concentrate on practical details of the schedule of emission goals and the design and implementation of policies to pursue them, without having to address broad, contentious, and potentially unresolvable questions of global equity. While early participants would bear higher mitigation costs with narrower participation, these costs may well be perceived as less objectionable than smaller costs that take the form of payments to other countries with extra emission credits to sell. These costs would also to a substantial extent take the form of investments in new technologies and expertise, which could have substantial commercial value subsequently as the mitigation regime expands. The approach would powerfully demonstrate the participating countries' commitment to take responsibility for their historical contribution to climate change, by leading the creation of a serious mitigation regime and accepting the costs of doing so. This is how it would justify, morally and politically, excluding nations unwilling to undertake initial serious mitigation commitments from negotiation of the details of the regime.

Whatever group of nations participates in the mitigation regime initially, the regime cannot stay limited to these participants if it is to be effective. Although starting with a few willing participants carries many practical advantages, it is essential that the initial agreement must promote, not obstruct, its own subsequent expansion. This imposes two requirements on the initial policies. They must limit the incentives for high-emitting industries to move to countries outside the agreement; and they must create the incentives for additional countries to join. To meet these two requirements, the coalition's initial mitigation agreement must include trade measures, to roughly equalize the cost burden from mitigation policies between internationally traded and domestically produced goods. Depending on the form of the mitigation policy adopted by the coalition, the required trade measures could take two principal forms. The first would be a border-tax adjustment, which would charge a tax on the emissions represented by imported products at the same rate as the cost per unit emission borne by domestically produced products, and rebate the mitigation cost to domestically produced goods being exported. The effect of a border-tax adjustment is to equalize the cost burden of the mitigation policy between equivalent goods produced inside and outside the coalition, wherever these are sold. An alternative measure with roughly the same effect would be to require imports to purchase emission permits as they enter the coalition, and to grant transferable emission permits to exports as they exit the coalition. The quantity of permits required or granted would be set to approximate the emissions that were generated in manufacturing the product. The calculation of border adjustments would have to be accurate enough in attributing emissions to imported products, accounting for both the energy system of the exporting country and the production technology of the traded goods. This would pose a serious challenge to data and administration systems, but probably not an insurmountable one. These trade measures would also have to be judged acceptable under the rules of the World Trade Organization (WTO). Their legality has not yet been precisely tested, although their prospects appear substantially more promising following a series of crucial recent WTO decisions (see, for example, Howse, 2002). Border adjustments for both exports and imports would be set at a lower level for trade with developing countries, perhaps initially at zero, than for trade with non-participating industrialized countries.

Even if an initially narrow mitigation regime is constructed with the right incentives to facilitate subsequent expansion, starting narrowly risks missing the opportunity to shift the rapid build-up of investment now underway in major developing countries, especially China, toward more advanced low-emitting technologies. This is a risk of this approach, albeit one it shares with every other proposal we have discussed except the bilateral USA–China approach. Indeed, it is possible that this approach would do better at exploiting the opportunity than any of the other proposals we have discussed – and it is not incompatible with the USA and China also pursuing a bilateral arrangement. Quick establishment of even a narrow rich-country coalition for serious mitigation would signal rich countries' commitment to address the issue more credibly than the present stalling and squabbling do, and would also immediately create some incentives for non-participating countries to lower their emissions. The proposed border measures would create such incentives immediately in export-oriented non-participating countries and investors in them, so long as the calculation of adjustments is sufficiently accurate and fair for imports entering the regime.

Creating incentives for initially non-participating countries to join, particularly developing countries, will be a crucial element of the approach, which will be more delicate than the superficially similar problem that was addressed in the early days of the Montreal Protocol. In that case, the risk of developing a parallel world economy outside the ozone treaty, producing and trading ozone-depleting chemicals and associated products among themselves, was occasionally mentioned but was never credible. But given the greater economic force of the developing countries today and the greater economic stakes in greenhouse-gas mitigation, the risk of a badly designed initial mitigation regime creating a parallel, outside coalition producing and trading with old, high-emitting technology is quite plausible and must be guarded against vigilantly. Initial design of an international mitigation regime must avoid creating such a hard-to-reverse split in the world economy.

A primary requirement for avoiding such a split will be addressing developing country governments' concerns that if they are not involved in negotiation of the initial mitigation strategy, the details of the strategy or the terms on which they are subsequently able to join might be biased against them. Two concerns are likely to be most important. The first is the risk that the border-tax measures will be too strict, conferring trading advantages on coalition producers rather than merely neutralizing the disadvantages imposed by their mitigation policies. This is a serious concern, although several factors would help to diminish it. First, since the adjustments will not be set by any single country but jointly by the whole coalition, divergent trade interests among them (i.e. which industries they would wish to favor, and by how much) will help restrain attempts to distort the adjustments for trade measures. Since the measures must be set multi-laterally, they will require some multi-lateral expert body to conduct the analysis and recommend the levels. This body could be designed to provide some insulation against political interference and provisions for appeal of its decisions. Moreover, the possibility of WTO challenge should discipline any attempt to use these measures to provide disguised trade advantages.

Developing countries' second major concern will likely be the terms on which they are able to join the mitigation coalition – their obligations to limit or reduce emissions, and any accompanying provisions for assistance in controlling their emissions. These negotiations will be complex and difficult, but similarly difficult negotiations would be required under any approach to developing a global mitigation regime. Relative to the other proposed approaches, this one has two advantages: that initial mitigation action need not await agreement on the terms of developing-country participation, and that the design of the initial mitigation strategy gives substantial incentives for additional countries, both industrialized and developing, to join. The most promising approach to negotiating developing countries' accession would resemble several of the proposals discussed above:

negotiating developing countries' accession to the regime relatively early, with emission limits somewhat above their projected emissions growth paths for the next 10 to 20 years, but which begin slowing emissions growth and then declining another 10 to 20 years thereafter. The specific terms could be negotiated so developing countries carry only a small share of the burden, or even receive a net benefit. Since flexible international shifting of mitigation effort among participating nations through voluntary transactions would be allowed, this approach would also allow international planning and negotiation for emission-reducing projects in developing countries to begin early, with clear accounting for their aggregate effects on emissions once national baselines were established.

This approach starts entirely outside the Kyoto Protocol. The coalition's initial approach could borrow elements from the Protocol, such as the design of flexibility mechanisms, but they cannot act within the Protocol because they need full control over negotiation of their initial mitigation commitments. But like the other proposals for parallel activity to advance mitigation outside the Protocol, this approach would also aim for eventual merger with the Protocol to create a single, comprehensive global climate-change regime.

This proposal is extremely challenging. It demands much political courage in the initial coalition of leading nations, and its subsequent development will require challenging negotiations between the initial coalition and the developing countries. Most difficult of all will be its implications for Europe–USA relations, particularly if political sentiment in the USA remains strongly opposed to significant mitigation efforts. Yet of all the approaches that have been proposed, we judge this one to have the greatest promise of success. It builds on existing evidence of political commitment, rather than assuming a large-scale change of heart on the part of nations presently resisting joining; it allows an orderly negotiation of the terms of an initial mitigation regime, under the control of those actually taking on commitments; it limits the risks borne by these early movers through trade measures that offset the competitive disadvantages they would otherwise suffer; and it provides a feasible path for the required expansion to global participation.

5.4.4   Adjusting responses over time

The final element required of a mitigation strategy is a procedure for reassessing and adjusting efforts over time. Although mitigation efforts must begin despite present uncertainty, the presence of uncertainty means that mitigation policies cannot be established once and for all. Expanding toward global participation is one dimension which a mitigation regime must adopt over time, but it is the simplest and most foreseeable dimension. In addition, the form and stringency of policies, the mix of technologies being developed and adopted, and even the long-term goal for climate stabilization, will all have to be repeatedly re-assessed and potentially revised over the many decades it will take to stabilize the climate. Many types of future changes in knowledge or capabilities may call for changes in these choices, including new scientific knowledge about the climate's sensitivity and speed of response to human forcing, the nature and severity of climate-change impacts, and the possibility of abrupt changes – as well as changes in technological capabilities to reduce emissions, new evidence on the effectiveness of policies, and other changes in relevant social and political conditions. In general, evidence of higher climate sensitivity, faster climate change, more severe impacts, or lower mitigation costs will call for strengthening mitigation efforts, despite the long lags between such efforts and their climatic effects. Conversely, evidence of lower climate sensitivity, slower changes, less severe impacts, or higher mitigation costs will suggest a decrease in the intensity of mitigation efforts.

It is not possible to anticipate what form changes in future knowledge or capabilities might take, so the details of how to adjust future efforts cannot be negotiated in advance. Rather, some future decision-making bodies will have to be given authority to assess changes and adjust policies in view of some agreed enduring principles or criteria. The outline of such a process for review and adjustment of commitments already exists in the Framework Convention, and a similar process has been used to great effect in the Montreal Protocol. As a more detailed and challenging set of mitigation commitments is developed, the process for reviewing and adjusting these over time will of course require further elaboration.

The most serious challenge for such a process will be balancing the need for policies to respond flexibly to new knowledge and capability with the need for a stable and credible policy trajectory to allow orderly investment and planning. There are various ways to balance these two priorities. For example, mitigation policies might be adopted as rolling long-term plans, with any significant adjustments being phased in gradually over periods of 5–10 years or longer. In addition, the disruptions from adjusting mitigation policies could be spread across the economy at minimum cost if they were implemented through market-based operations. In such operations, governments would change the availability of emission permits for some future year by buying back permits on the open market to decrease the supply, or by auctioning additional ones to increase the supply.

5.5   Conclusion

In this book, we have summarized present scientific knowledge about how and why the climate is changing, how it is likely to change over the coming century, what the associated impacts might be, and what can be done about it. Our conclusion is that scientific knowledge about present and likely future climate changes calls for an urgent, high-priority response – principally but not exclusively through international negotiation of coordinated national policies – to reduce future emissions and to prepare for a much more uncertain and potentially less benign climate than we have been fortunate to live in for the past century. Concrete efforts to construct such a response must begin immediately.

But we do not yet have a serious response. There are many reasons for this. Some are related to the intrinsic difficulty of the issue, which challenges our present decision-making systems. Some are related to the inevitability of scientific uncertainty – which does not justify a stance of inaction, but which does provide rhetorical opportunities for opponents of action to confuse the issue and advocate delay. Whatever the mix of reasons, the present policy response is utterly inadequate in view of the gravity of the climate-change issue. A few nations are approaching the starting line of taking the issue seriously, but most are not even close. The state of international decision-making, where the main action must occur, is ineffective, incoherent, and deadlocked.

In view of the present grave situation, the previous section has sketched and briefly assessed the major alternatives proposed to the present approach. While many of these appear unpromising, two appear to hold some prospect of success: a USA–China bilateral agreement; and more promisingly, an industrialized-country "coalition of the willing" taking on significantly stronger mitigation goals and measures, and adopting trade measures that would both reduce their resultant competitive disadvantage and give other nations incentives to join them. These alternatives, including the one we judge most promising, were presented as sketches rather than detailed policy proposals. They were intended to make the case that movement toward a serious mitigation regime with commitments to real, long-term emission reductions, is not just essential to forestall serious future climatic risks, but is also practically and politically feasible.

More important than the precise details of initial mitigation policies is the structure of continuing research, periodic assessment, and review of policies and goals through which they are progressively adapted over time as knowledge and capabilities advance. Over time, relevant uncertainties – about climate change, impacts, and options to adapt or reduce emissions – can be reduced through sustained programs of research, development, and assessment, although not eliminated. Policies should be designed to pursue complementarities and multiple benefits – in terms of harnessing positive feedbacks in innovation, and in terms of seeking directions of innovation that promise joint management of multiple environmental or other issues. We will have to continue to make decisions under uncertainty, and the details of policy will have to be worked out progressively through negotiation, experimentation, and review. At present, precious little is being done to pursue any of these seemingly reasonable and modest directions.

Getting to a climate-policy regime that will be sustainable, adaptable, and practical, depends on taking the first steps, even if our knowledge of where our ultimate destination lies is only approximate.

Managing human influences on the Earth's climate is like piloting a super-tanker through dangerous waters. We do not know for sure, but it looks increasingly likely that there are rocks ahead: in fact, we might be pointed right at one. We know what direction we need to steer, but do not know how far we must steer to avoid this rock, whether there are other rocks around, or how hard we can steer without risking damage to the ship. Moreover, a big ship like this one takes miles to change course. Unfortunately, no one is at the wheel right now. The crew is downstairs, arguing about whether there really are rocks ahead, what the precise course is that we must steer to reach our ultimate destination, and whose job it is to steer. While the crew is arguing, the ship is getting closer to the rocks. Somehow, what we need is to get someone upstairs to start steering us away from the rocks – now. Because the steering is so slow, it must start right away. At the same time, we need to learn more about where the rocks are – and also to learn, by starting to steer, about how the ship responds and how hard we can steer it. But neither of these needs to learn more justifies waiting to start the steering: they just mean we must steer very carefully, and be vigilant to everything we can learn about the ship and the hazards in the waters, while we do it. We can probably avoid the rocks, but we need to start now.

Further reading for Chapter 5

Aldy, J. E., Ashton, J., Baron, R., Bodansky, D., Charnovitz, S., Diringer, E., Heller, T., Pershing, J., Shukla, P. R., Tubiana, L., Tudela, F., and Wang, X. (2003). *Beyond Kyoto: Advancing the International Effort Against Climate Change*. Washington, DC: Pew Center on Global Climate Change, December.

> A collection of six essays examining specific aspects of a potential international climate-change regime, including long-term targets; near-term commitments, international equity, costs, and the connections of climate-change policy to economic development and international trade.

Aspen Institute (2002). *U.S. Policy on Climate Change: What's Next?* A report of the Aspen Institute Environmental Policy Forum, Frank Loy and Bruce Smart (co-chairs), ed. John A. Riggs. Aspen, Colorado: Aspen Institute.

> The results of a senior bipartisan forum convened by the Aspen Institute in 2002. In addition to the chairs' summary of the major conclusions of the forum, the report includes brief background papers that review major areas of the current policy debate about emission trends, technologies, costs, and potential policy responses.

Howse, R. (2002). The Appellate Body Rulings in the Shrimp/Turtle Case: a new legal baseline for the trade and environment debate. *Columbia Journal of Environmental Law*, **27**(2), 489–519.

A discussion of crucial recent WTO rulings on the US ban on imports of shrimp harvested by nations that do not match the US policy requiring turtle-excluder devices. Although the initial US policy was rejected for being discriminatory in its application, the decision greatly strengthened the ability of environmental measures that are not discriminatory to use trade restrictions in pursuit of environmental objectives.

Parson, E. A. (2003). *Protecting the Ozone Layer: Science and Strategy*. New York: Oxford University Press.

This history of the interwoven progression of scientific, technological, and political debates concerned with depletion of the stratospheric ozone layer identifies several central lessons from the failures and successes of the ozone regime that can be applied to help break the present policy deadlock on global climate change.

Rowland, F. S. (1993). President's Lecture: The Need for Scientific Communication with the Public. *Science*, **260** (11 June), 1571–1576.

In this President's Address to the 1993 annual meeting of the American Association for the Advancement of Science, Rowland reviews some of the pseudo-scientific claims about ozone depletion then circulating in popular and policy settings, notes how easy it is to make such claims appear persuasive to a non-scientific audience, and argues the need for greater scientific education of the public and policy-makers – including greater education for skeptical examination of scientific claims advanced in policy settings.

Sandalow, D. B. and Bowles, I. A. (2001). Fundamentals of treaty-making on climate change. *Science*, **292** (8 June), 1839–1840.

A brief summary of the status of international climate policy after the Bush Administration's rejection of the Kyoto Protocol, and a discussion of those aspects of international policy that are widely accepted as necessary elements of a resolution of the issue.

Stewart, R. B. and Wiener, J. B. (2003). *Reconstructing Climate Policy: Beyond Kyoto*. Washington, D.C.: American Enterprise Institute.

This monograph argues that it is in America's national interest to take a more active stance on climate change, and proposes a path forward based on bilateral USA–China negotiations of joint emission limits and a well managed emission trading system, which could be subsequently expanded by the accession of additional countries and would eventually aim at merging with the Kyoto Protocol.

Taubes, G. (1993). The ozone backlash. *Science*, **260** (11 June), 1580–1583.

A news article by a staff writer of *Science* magazine provides greater detail on the events and specific claims advanced in the "ozone skeptics" backlash of the early 1990s. Best read in conjunction with Rowland's presidential address, cited above, which appears in the same issue of *Science*.

Victor, D. G. (2004). *Climate Change: Debating America's Policy Options*. New York: Council on Foreign
    Relations.

> In addition to general background on climate-change science and policy, this briefing
> note includes sketches of three alternative paths for US climate policy: a relatively
> passive response that relies on adaptation and technological change to manage the
> issue; an attempt to develop global emission-permit markets independent of
> negotiation of emission limits; and an attempt to re-engage the Kyoto Protocol
> process and take the lead in addressing its weaknesses.

Appendix

A1   Present value and discounting

We state in Chapter 4 that costs incurred in the future have a lower "present value." What does this mean? The present value is the cost *today* of some future expense. One can think of the present value of an expense as the amount of money you need to invest today so that you can pay the cost when it is incurred. For example, if you know you will incur a cost of $100 in 10 years, you could invest $50 today at 7% interest rate in order to have $100 in 10 years, when your cost occurs. In this case, we'd say the present value of the $100 cost is $50. Implicit in any discussion of present value is an interest rate, which is usually referred to as the "discount rate." It is the rate of return of the invested money, and is usually a few percent. Changing the discount rate can greatly affect the present value.

Mathematically, one can calculate the present value of a cost incurred sometime in the future as:

$$PV = \frac{cost}{(1+r)^n} \tag{A1.1}$$

where *PV* is the present value, *cost* is the amount of the expected expense ($100 in the previous example), *r* is the discount rate (7% in the previous example, but expressed in the equation as 0.07), and *n* is the number of years until the expense is incurred (10 years in the previous example).

Given a fixed cost, as the length of time before the cost is incurred increases, the smaller amount you need to invest. In other words, the present value of a cost decreases as the expense recedes further into the future. If you had 20 years before you incurred the same $100 cost, for example, you'd have to invest only $25 today at 7% interest. Thus, when calculating the cost of various climate change regulation scenarios, scenarios that defer the costs furthest into the future will generally have the lowest present-value costs.

The advantage of the present-value concept is that it allows you to express all costs on a common scale, so they can be compared. In this way, it is possible to determine which of several different scenarios, each with a different schedule of costs over the next century, is the cheapest.

However, there are some problems with the concept. Exponential discounting expressed by Equation (A1.1) tends to reduce costs that are many decades in the future to near zero today. For example, a $100 cost that is to be incurred in 100 years has a present value of only 11 cents. For such long time horizons, there are reasons to believe that exponential discounting underestimates the true present value of future expenses.

A2   Marginal costs

Consider a plant that emits 100 tons per year of some pollutant. Reducing emissions to 99 tons per year is relatively easy and costs little. No new equipment might be required; perhaps the equipment in the plant can be tuned up, or the operational procedures modified. Reducing the emissions from 99 to 98 tons per year takes a little more effort than the first ton, and therefore costs a little more. The cost of reducing the emissions from 98 to 97 tons per year costs even more. And so on.

The marginal cost of some action is the cost of an incremental change in the level of the action (or in calculus terms, the partial derivative of total cost with respect to changes in the level of action). In this example, the marginal is the cost of reducing each additional ton of pollutant. In controlling emissions of a pollutant, if emissions are being reduced by 100 tons, the marginal cost of this policy is the cost of the hundredth ton reduced – i.e. the additional cost of going from cutting 99 tons to cutting 100 tons. As a pollutant is controlled more and more tightly, the marginal cost almost always increases (i.e. the second derivative is positive).

One can also talk about marginal benefits. The marginal benefit is the benefit obtained from reducing that last ton of pollutant. In general, the marginal benefit decreases as the pollutant is controlled more and more tightly.

As we discuss in Chapter 4, policy decisions can be made by comparing marginal costs and marginal benefits. In general, the optimal policy is set when a pollutant is cut to the point where marginal costs and marginal benefits are equal to each other.

A3   A quantitative example of emissions permit trading

To understand the advantages and disadvantages of flexible mitigation strategies, consider the following hypothetical example. Imagine two plants, A and B, both of which produce 1000 units of some pollutant. The marginal cost of reducing emissions in plant A is x. This means that plant A can reduce their emissions by one unit for $1, reducing one more unit costs an additional $2, another unit costs an additional $3, etc. Cutting pollution by three units at plant A therefore costs $1 + $2 + $3 = $6. Plant B is older and contains less technically advanced equipment, and therefore its marginal cost of reducing emissions is $2x$, twice the cost of plant A (and its total cost is twice, also).

Under a conventional regulatory approach, both plants are required to reduce their pollution by 10 units. It costs plant A $55 ($1 + $2 + ⋯ + $9 + $10) to do this, while it costs plant B $110. The total cost to the economy to reduce pollution emitted to the atmosphere by 20 units is $165.

It turns out that there are cheaper and more equitable ways to achieve the same reduction. One such way is a tradable-permit system. The government issues each plant permits to emit 990 units, ten fewer than they are presently emitting. Imagine also that a market exists for these emissions permits, and the market value of 1 permit is $14. Both plant A and plant B will cut to the point where their marginal cost is equal to the value of the permit. Plant A will cut 14 units of emissions – so its total emissions are 986. Since it was issued 990 permits, it will have 4 unused permits and these can be sold on the open market for $14 each. The net cost to plant A of complying with the emissions regulation is therefore $49 (cost of reducing emission to 986 is $105, minus revenue from selling the extra permits, $14 × 4 = $56). Plant B will cut 7 units of emissions – so its total emissions are 993 and it will have to buy 3 more permits on the open market. The total cost to plant B of complying with the emissions regulation is $98 (cost of reducing emission to 993 is $56, plus the cost of buying permits, $14 × 3 = $42). The total cost to the economy is $147 for a reduction of 21 units.

One way to think about this is that plant B has paid plant A to make some of plant B's reductions. This makes sense for both of the plants because the amount plant B paid to plant A was less than the amount plant B would have paid to make those reductions themselves, and more than it cost plant A to make the reductions. As a result, the cost of complying with the regulations is less for both plants than under a conventional regulation. And the total cost to the economy is less.

If the permits are exchanged on an open market, then the value of the permits would go up and down until an appropriate price is reached. The number of permits issued would set the total emissions to the atmosphere, and the reduction in emissions would have been obtained at a lower cost than under a convectional regulatory approach. In order to achieve the long-term emissions reductions necessary to curtail climate change, the number of permits issued to the government would decrease in time according to a schedule known long in advance.

Now consider a tax on each unit of pollution emitted rather than a permit system. Under such a tax, both plants A and B would reduce emissions until the marginal cost of reduction is equal to the tax. If the tax is set at $14 per unit of pollution, then the two plants will make exactly the same reductions as under the permit plan. In fact, the reductions expected by a tax are the same as by a permit system when the price of the permit is equal to the tax.

Glossary

adaptation   Reacting to the changes in the climate. For example, if sea level rises, adaptation measures might include building a seawall or relocating people who live near the ocean farther inland.

aerosols   Small solid or liquid particles suspended in the atmosphere, including dust and soot. The net impact of these particles on the climate is not currently well understood.

cap and trade   A regulatory system in which permits are distributed that allow holders to emit a specified amount of greenhouse gas to the atmosphere. The total number of permits therefore defines the total amount of greenhouse gases emitted (the "cap"). The permits can be traded, allowing them to be used by the emitters with the highest marginal costs. See Appendix A3 for an example of how this works.

CH_4   *See* methane

climate sensitivity   The change in the Earth's climate caused by a specified change in CO_2. In most cases, the climate sensitivity is the eventual warming that occurs when the pre-industrial atmospheric concentration of CO_2, 270 p.p.m.v., is suddenly doubled to 540 p.p.m.v., then held at that higher level forever. Our most recent estimates put this doubled-CO_2 sensitivity at 1.5 °C to 4.5 °C.

CO_2   *See* carbon dioxide

CO_2-equivalent   The amount of CO_2 that would cause the same amount of global warming as a given mixture of CO_2 and other greenhouse gases.

carbon dioxide (CO_2)   Greenhouse gas produced during combustion of fossil fuels or when biomass is burned. Its present abundance in the atmosphere is about 375 p.p.m.v., while before the industrial revolution it was about 270 p.p.m.v.

deforestation   The process of clearing land of forests. Usually, the trees are burned and the carbon contained in them is released to the atmosphere, increasing atmospheric

CO_2. Recent estimates suggest that deforestation contributed about 1.6 GtC of carbon to the atmosphere in 2000, out of a total human contribution that year of about 8 GtC.

FCCC   *See* Framework Convention on Climate Change

Framework Convention on Climate Change (FCCC)   The first international treaty on climate change, it was signed in June 1992 and entered into force in 1994. It has since been established law in all the nations that have ratified – now numbering nearly 190, including the United States of America. The FCCC contains few binding requirements, but was rather intended to provide a structure within which more specific and binding measures could be negotiated later. Importantly, the treaty includes the concepts of "common but differentiated responsibilities" and keeping greenhouse gases below levels that are dangerous.

GCM   *See* general circulation model

GDP   *See* gross domestic product

general circulation model   Computer programs that use the known physics governing the Earth to simulate the state of the climate. These can be used to examine causes of past variations in the climate or to predict how various policies will affect the future state of the climate.

geoengineering   Actively manipulating the climate to offset the effects of increased greenhouse gases in the atmosphere. An example is launching a sunshade into space to shade the Earth.

GtC   Gigatons of carbon. A gigaton is equal to 1 billion metric tons, and a metric ton is equal to 1000 kg or 2200 lbs. When this unit is applied to emissions, usually only the mass of carbon is counted (thus ignoring the mass of oxygen).

gross domestic product (GDP)   The total value of goods and services produced by an economy. Per capita GDP (GDP divided by the population) is a measure of the wealth or affluence of the society.

Intergovernmental Panel on Climate Change (IPCC)   Established by the World Meteorological Organization (WMO) and the United Nations Environment Programme (UNEP), the role of the IPCC is to review and evaluate the peer-reviewed literature on the science of climate change in order to determine the areas in which there exists a consensus and which areas there does not. The IPCC publishes reports on the status of the scientific community's understanding of climate change every five years.

internal variability   Changes in the climate that occur without any external forcing factor like changes in the amount of sunlight. The most familiar example of internal variability is the Southern Oscillation, which comprises the El Niño/La Niña duo.

IPCC   Intergovernmental Panel on Climate Change.

marginal cost   The marginal cost of some action is the cost of an incremental change in the level of the action (or in calculus terms, the partial derivative of total cost with respect to changes in the level of action). See Appendix A2 for a discussion of this concept.

methane (CH_4)   This is an important greenhouse gas, which is emitted from rice paddies, landfills, livestock, and the extraction and processing of fossil fuels, as well as several natural sources. While emitted in much smaller quantities than CO_2, it contributes substantially more warming per pound emitted, so it plays an important role in the climate change problem.

metric ton   1000 kg or 2200 lbs.

mitigation   Reducing emissions of CO_2 and other greenhouse gases so that the climate never changes in the first place.

N_2O   *See* nitrous oxide.

nitrous oxide (N_2O)   This is an important greenhouse gas, which is emitted from natural as well as various agricultural and industrial processes. While emitted in much smaller quantities than CO_2, it contributes substantially more warming per pound emitted, so it nonetheless plays an important role in the climate change problem.

parts per million (p.p.m.)   This is a unit for expressing the abundance of trace gases in the atmosphere. An abundance of 1 p.p.m. means that there is one molecule of the gas of interest in every million molecules of air. Today's atmospheric CO_2 abundance is 380 p.p.m.v., meaning that 380 out of every million molecules in the air are CO_2.

proxy climate record   A proxy climate record is a record of past climate variation that has been imprinted on some long-lived physical, chemical, or biological system. Because of their longevity, climate proxies can provide evidence of past climate from long before the modern instrumental record. Climate proxies include tree rings, ice cores, corals, ocean sediments, and boreholes.

scientific assessment   A report generated by a group of scientists that summarizes important findings of the scientific community on questions of relevance to policymakers. For the climate arena, the main assessment body is the IPCC.

Working Group I   A subgroup of the IPCC focused on the atmospheric science of climate change.

Working Group II   A subgroup of the IPCC focused on the impacts of climate change and potential for adapting to the changes.

Working Group III   A subgroup of the IPCC focused on the potential to reduce the greenhouse-gas emissions contributing to climate change.

References

Aldy, J. E., Ashton, J., Baron, R., Bodansky, D., Charnovitz, S., Diringer, E., Heller, T., Pershing, J., Shukla, P. R., Tubiana, L., Tudela, F. and Wang, X. (2003). *Beyond Kyoto: Advancing the International Effort Against Climate Change*. Washington, DC: Pew Center on Global Climate Change, December.

Aspen Institute (2002). *U.S. Policy on Climate Change: What's Next?* A report of the Aspen Institute Environmental Policy Forum, Frank Loy and Bruce Smart (co-chairs), ed. John A. Riggs. Aspen, Colorado: Aspen Institute.

Bell, T. L., Chou, M.-D., Hou, A. Y. and Lindzen, R. S. (2002). Reply. *Bull. Am. Met. Soc.*, **83**, 599.

Bimber, B. (1996). *The Politics of Expertise in Congress: the Rise and Fall of the Office of Technology Assessment*. Albany: SUNY Press.

Caldeira, K., Jain, A. K. and Hoffert, M. I. (2003). Climate sensitivity uncertainty and the need for energy without CO_2 emission. *Science*, **299**, 2052–2054.

Chou, M.-D., Lindzen, R. S. and Hou, A. Y. (2002). Comments on "The Iris hypothesis: a negative or positive cloud feedback?". *J. Climate*, **15**, 2713–2715.

Christy, J., Spencer, R. W., Norris, W. B., Braswell, W. D. and Parker, D. E. (2003). Error estimates of version 5.0 of MSU-AMSU bulk atmospheric temperatures. *J. Atmos. Oceanic Technol*, **20**, 613–629.

Dessler, A. E. (2000). *The Chemistry and Physics of Stratospheric Ozone*. San Diego: Academic Press.

Fu, Q. *et al.*, (2004). Contribution of stratospheric cooling to satellite-inferred tropospheric temperature trends. *Nature*, **429**, 55–58.

Harrison, H. (2002). Comment on "Does the Earth have an adaptive infrared Iris?". *Bull. Am. Met. Soc.*, **83**, 597.

Hartmann, D. L. & Michelsen, M. L. (2002). No evidence for Iris. *Bull. Am. Met. Soc.*, **83**, 249–254.

Hoffert, M. I. (2002). Advanced technology paths to global climate stability: energy for a greenhouse planet. *Science*, **295**, 981–987.

Howse, R. (2002). The Appellate Body Rulings in the Shrimp/Turtle Case: a new legal baseline for the trade and environment debate. *Columbia Journal of Environmental Law*, **27** (2), 489–519.

IPCC (1996). *Climate Change 1995: The Science of Climate Change. Contribution of Working Group I to the Second Assessment Report of the Intergovernmental Panel on Climate Change*, ed. J. T. Houghton,

L. G. Meira Filho, B. A. Callander, N. Harris, A. Kattenberg and K. Maskell. Cambridge and
New York: Cambridge University Press.

(2000). *Emissions Scenarios: Special report of the Intergovernmental Panel on Climate Change*, ed.
N. Nakicenovic and R. Swart. Cambridge and New York: Cambridge University Press.

(2001a). *Climate Change 2001: The Scientific Basis. Contribution of Working Group I to the Third
Assessment Report of the Intergovernmental Panel on Climate Change*, ed. J. T. Houghton, Y. Ding,
D. J. Griggs, M. Noguer, P. J. van der Linden, X. Dai, K. Maskell, and C. A. Johnson. Cambridge
and New York: Cambridge University Press.

(2001b). *Climate Change 2001: Impacts, Adaptation, and Vulnerability. Contribution of Working Group II
to the Third Assessment Report of the Intergovernmental Panel on Climate Change*, ed. J. J. McCarthy,
O. F. Canziani, N. A. Leary, D. J. Dokken and K. S. White. Cambridge and New York:
Cambridge University Press.

(2001c). *Climate Change 2001: Mitigation. Contribution of Working Group III to the Third Assessment
Report of the Intergovernmental Panel on Climate Change*, ed. B. Metz, O. Davidon, R. Swart and
J. Pan. Cambridge and New York: Cambridge University Press.

(2001d). *Climate Change 2001: Synthesis Report. A Contribution of Working Groups I, II, and III to the
Third Assessment Report of the Intergovernmental Panel on Climate Change*, ed. R. T. Watson and
the Core Writing Team. Cambridge and New York: Cambridge University Press.

Jasanoff, S. (1990). *The Fifth Branch: Science Advisors as Policymakers*. Cambridge, MA: Harvard
University Press.

Karl, T. R. and Trenberth, K. E. (2003). Modern global climate change. *Science*, **302** (5 December),
1719–1723.

Kuhn, T. S. (1962). *The Structure of Scientific Revolutions*. Chicago: University of Chicago Press.

Lean, J. and Rind, D. (2001). Sun–climate connections: Earth's response to a variable Sun. *Science*,
292, 234–236.

Lin, B., Wong, T., Wielicki, B. A. and Hu, Y. (2004). Examination of the decadal tropical mean
ERBS nonscanner radiation data for the Iris hypothesis. *J. Climate*, **17**, 1239–1246.

Lindzen, R. S., Chou, M.-D. and Hou, A. Y. (2001). Does the Earth have an adaptive iris? *Bull. Am.
Met. Soc.*, **82**, 417–432.

(2002). Comment on "No evidence for Iris". *Bull. Am. Met. Soc.*, **83**, 1345–1348.

Lomborg, B. (2001). *The Skeptical Environmentalist: Measuring the Real State of the World*. New York:
Cambridge University Press.

Mazur, A. (1973). "Disputes between experts". *Minerva: a review of science, learning, and policy*, **11**: 2,
April 1973, pp. 243–262.

McIntyre, S. and McKitrick, R. (2005). Hockey sticks, principal components, and spurious
significance. *Geophys. Res. Letts.*, **32** (3), L03710.

Mears, C. A. *et al.* (2003). A reanalysis of the MSU channel 2 tropospheric temperature trend. *J.
Climate*, **16**, 3650–3664.

Michaels, P. J. (2004). *Meltdown: The Predictable Distortion of Global Warming by Scientists, Politicians,
and the Media*. Washington, D.C.: Cato Institute.

Michaels, P. J. and Balling, R. C. (2000). *The Satanic Gases: Clearing the Air About Global Warming*.
Washington, D.C.: Cato Institute.

Moberg, A., Sonechkin, D. M., Holmgren, K., Datsenko, N. M. and Karlén, W. (2005). Highly
variable Northern Hemisphere temperatures reconstructed from low- and high-resolution
proxy data. *Nature*, **433**, 613–617.

Monastersky, R. (2003). Storm brews over global warming. *Chronicle of Higher Education*, **50** (2), A16.

National Commission on Energy Policy (2004). *Ending the Energy Stalemate: A Bipartisan Strategy to Meet America's Energy Challenges*. December 2004. Available at www.energycommission.org

Oerlemans, J. (2005). Extracting a climate signal from 169 glacier records. *SciencExpress*, 3 March, 10.1126/science.1107046.

Pacala, S. and Socolow, R. (2004). Stabilization wedges: solving the climate problem for the next 50 years with current technologies. *Science*, **305**, 968–972.

Parson, E. A. (2003). *Protecting the Ozone Layer: Science and Strategy*. New York: Oxford University Press.

Parson, E. A. and Fisher-Vanden, K. (1997). Integrated assessment models of global climate change. *Annual Review of Energy and the Environment*, **22**, 589–628.

Petit, J. R. *et al.* (1999). Climate and atmospheric history of the past 420,000 years from the Vostok ice core, Antarctica. *Nature*, **399**, 429–436.

Ruckelshaus, W. D. (1985). Risk, science, and democracy. *Issues in Science and Technology*, **1**: 3, Spring, pp. 19–38.

Sandalow, D. B. and Bowles, I. A. (2001). Fundamentals of treaty-making on climate change. *Science*, **292** (8 June), 1839–1840.

Stewart, R. B. and Wiener, J. B. (2003). *Reconstructing Climate Policy: Beyond Kyoto*. Washington, DC: American Enterprise Institute.

Soon, W. and Baliunas, S. (2003). Proxy climatic and environmental changes of the past 1000 years. *Climate Research*, **23**, 89–110.

Spencer, R. W. and Christy, J. R. (1990). Precise monitoring of global temperature trends from satellites. *Science*, **247**, 1558–1562.

Stommel, H. M. and Stommel, E. (1983). *Volcano Weather: the Story of 1816, the Year without a Summer*. Newport, RI: Seven Seas Press.

von Storch, H., Zorita, E., Jones, J. M., Dimitriev, Y., Gonzalez-Rouco, F. and Tett, S. F. B. (2004). Reconstructing past climate from noisy data. *Science*, **306** (5296), 679–682.

US National Academy of Sciences (2000). Commission on Geosciences, Environment and Resources. *Reconciling Observations of Global Temperature Change*. Washington, D.C.: National Academy Press.

(2003). Climate Research Committee, Panel on Climate Change Feedbacks. *Understanding Climate Change Feedbacks*. Washington, D.C.: National Academy Press.

US Global Change Research Program, National Assessment Synthesis Team (2001). *Climate Change Impacts on the United States: The Potential Consequences of Climate Variability and Change*. New York: Cambridge University Press.

Victor, D. G. (2004). *Climate Change: Debating America's Policy Options*. New York: Council on Foreign Relations.

Vinnikov, K. Y. and Grody, N. C. (2004). Global warming trend of mean tropospheric temperatures observed by satellites. *Science*, **302**, 269–272.

Weart, S. R. (2003). *The Discovery of Global Warming*. Cambridge, MA: Harvard University Press.

Weinberg, A. M. (1972). "Science and trans-science". *Minerva: a review of science, learning, and policy*, **10**: 2, April 1972, pp. 209–222.

Index